VOICES IN THE CANYON

by
Catherine W. Viele

Library of Congress Catalog Card No.: 78-68825
ISBN No.: 911408-52-5

Book Design — Christina Watkins Fall
Production Specialist — Bob Petersen
Typography — Tiger Typographics, Flagstaff
Printing — Graphic Dimensions

CONTENTS

VOICES IN TIME

"The time weathered and silent remains of this long departed people hint at a way of life deeply rooted in the natural world, and challenge our intellect and imagination to uncover their hidden story."

Deep in the sandstone canyons of northern Arizona, even today one of the most remote areas of the country, lie the deserted villages of an ancient people who, by hard work and ingenuity, carved out a living in a ruggedly beautiful but inhospitable land. These people, the Anasazi, nestled their homes in the rock overhangs, fashioning them of the same red rock in which they were set. Out of these storied pueblo towns, they climbed down to till their fields on the canyon bottoms below. Today, the time-weathered and silent remains of this long departed people hint at a way of life deeply rooted in the natural world, and challenge our intellect and imagination to uncover their hidden story.

Navajo National Monument preserves three of the most impressive of these Anasazi cliff dwellings — Betatakin, set in a sandstone arch of such magnitude that it dwarfs the village within; Keet Seel, the largest cliff dwelling in Arizona, but especially remarkable for its exceptional state of preservation; and Inscription House, a smaller cliff dwelling named after an inscription carved in historic times on the wall of one of its rooms. These sites, occupied during the last half of the 13th century, represent the peak of the Anasazi tradition in the area, a lifeway over 1000 years in development. Sometime around A.D. 1300, the people seem to have drifted away, leaving only their abandoned towns as testimony of what was once a vital culture and way of life.

Where are they today? Modern day Hopi Indians, living some 80 kilometers (50 miles) to the south on the edge of Black Mesa, claim the Anasazi of Betatakin, Keet Seel and Inscription House as their ancestors, and archeologists are generally in agreement. The Hopi are a deeply traditional people. The way of life that they have led for hundreds of years, perched high on the edge of precipitous mesas, is probably in many ways similar to that of their prehistoric ancestors. In fact, it is by ethnographic inference — making comparisons and assuming similarities between historic Hopi and prehistoric Anasazi — that archeologists have learned much of what they understand about the Anasazi.

Tales about the migrations of the various Hopi clans form an important part of Hopi religious tradition. The stories relate that after the emergence of the people into this world from the one beneath, the clans spent many years wandering before finally coming together to settle on the Hopi mesas, to them the center of the world. The Hopi believe that what they call "Kawestima" or "North Village" — the area around what we today call Betatakin, Keet Seel, and Inscription House — was one of the last places where some of the clans lived before final migration to the Hopi land. A large pictograph at Betatakin may provide evidence for this story; it

Betatakin — unlike Keet Seel and Inscription House — is a single occupation site. The 13th century Anasazi villagers were the first and last people to make use of the rock shelter on a permanent basis.

Inscription House, a smaller village located in a different canyon system, was occupied at the same time as Betatakin and Keet Seel.

has been tentatively identified as a representation of the Hopi Fire Clan symbol.

Today this area is in the middle of the vast Navajo Indian reservation. The Navajo are relatively recent arrivals to the Southwest; they wandered in from the north sometime around the 1300's. The Anasazi had by this time abandoned the cliff dwellings and moved to the south. The word "Anasazi" is, in fact, a Navajo one meaning the "Ancient Ones"—or, more exactly, "Ancestors of the Alien People." This is how the Navajo refer to the departed people whose deserted villages they found in their homeland.

It is because the surrounding land now belongs to the Navajo that the National Park Service has preserved the three cliff dwellings under the name Navajo National Monument. This is in spite of the fact that the Indians who built them—the Anasazi—were not Navajo but a Pueblo people.

The cliff dwellings that most people associate with the Anasazi actually represent only a small part of the total Anasazi story. Though Anasazi people

This 1896 shot of Walpi — a Hopi village on First Mesa — looks much as Keet Seel or Betatakin probably did when the Anasazi lived there.

were in the area for over 1000 years, they spent only the last 50 in the cliff villages. Most of this long time span they lived, not in the cliffs and canyons and not in large villages, but in mesa top homes and in smaller groups. In fact, even during the time that some of the Anasazi occupied cliff dwellings, probably half of the people were in smaller sites on the mesa or in villages out in the open. The common conception of the Anasazi as only cliff dwellers is misleading and incomplete.

The whole story of the Anasazi is far more fascinating and complex. During the time they occupied the area they evolved from simple hunters and gatherers into town dwellers with a sophistication and technology unknown before that time. The earliest Anasazi groups were small, wandering bands who lived off the land, collecting roots and berries in season and stalking wild game But with time they gradually learned the technology which radically altered their way of life — how to build masonry walls from materials close at hand; how to form clay into pottery that

"The Hopi are a deeply traditional people. The way of life that they have led, perched high on the edge of precipitous mesas, is probably in many ways similar to that of their prehistoric ancestors."

was both useful and decorative; how to plant seeds to produce a more dependable food supply.

Eventually, the people gathered together into larger groups, and towns like Betatakin and Keet Seel grew up. With this new "urban" way of life came further cultural sophistications — more complex social organization, greater specialization of labor, more elaborate

Anasazi ancestors of the Hopi Fire Clan may have painted this symbol at Betatakin.

Hopi symbol — whirlwind signifying the coming of rain.

religious traditions and ceremonies. The Anasazi had come a long way from their simple beginnings to attain a new and previously untried level of civilization.

What insights can we gain from the Anasazi? In their story we can see rudimentary beginnings for some of the same ways that form the basis of our own highly technological and urbanized society. Archeology is not just a catalogue of past events, practices, and materials; it is also an attempt to explain the past — to look for what is common to men of all ages and places, and to benefit from their experiences. The Anasazi story calls to mind many questions that are central to archeological inquiry, and may eventually help archeologists find some answers. Some of these questions our own civilization might do well to consider, and perhaps to learn from the Anasazi.

CANYON & MESA

Betatakin and Keet Seel ruins are set in drainages of the extensive Tsegi Canyon. The Tsegi is actually a great system of finger canyons carved into Skeleton Mesa and emptying at its mouth into Marsh Pass, between Longhouse Valley to the west and the western edge of the broad Chinle Valley to the east. Inscription House ruin is located in an arm of the complex Navajo Canyon system, a separate and even more widespread drainage west and north of the Tsegi. Both the Tsegi and Navajo systems are part of the vast network of canyons and mesas that fall away from the uplift of Navajo Mountain, the great rounded dome — sacred to the Navajo and Hopi Indians — which dominates the horizon in the area.

"Both the Tsegi and Navajo systems are part of the vast network of canyons and mesas that fall away from the uplift of Navajo Mountain, the great rounded dome — sacred to the Navajo and Hopi Indians — which dominates the horizon in the area."

The Colorado Plateau in northeastern Arizona is dissected by a myriad of such intricately carved canyons, sculpted in the soft sandstone by the slow processes of time and erosion. Water — flowing in spring-fed streams and rushing in torrents from late summer flash floods — is the main erosive agent; also important are the wearing forces of the winds, the scouring of wind or water borne sand particles, and the expansion and contraction of the rock with temperature changes. A general uplift of the whole plateau area over a period of many millions of years helped to hasten the downcutting process, increasing the speed and cutting

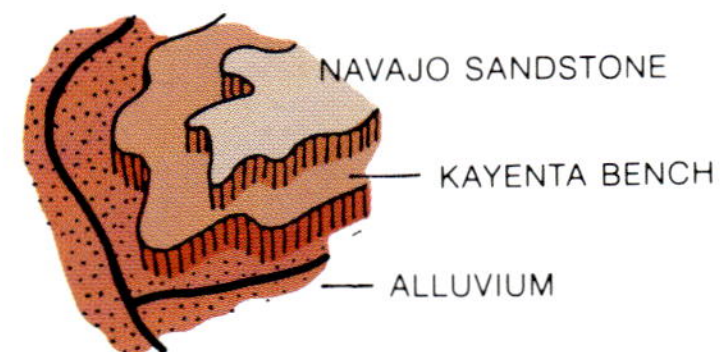

TSEGI CANYON SYSTEM

force of the waters in their flow to the sea.

Erosion has cut through and revealed several layers of sedimentary rock, representing about 40 million years of geologic history. The uppermost formation is the Navajo sandstone; it forms the "slickrock"—exposures of bare rock—on the mesa top and the towering red cliffs immediately below. This rock, some 180 million years old, is composed of layered sand dunes, solidified after millions of years under extreme pressure. Angled lines etched in the Navajo sandstone cliffs show the sloping sides of these ancient, wind-blown sand dunes. This cross-bedding, as it is called, and the weird erosional forms of the slickrock recall the sandy origin of the formation.

The thin, irregular benches of pale, purplish-red rock which start immediately below the Navajo sandstone cliffs are the Kayenta formation. Though also a sandstone, it is horizontally bedded with layers of mud and shale. Wingate sandstone has been exposed beneath the Kayenta in some deeper parts of the canyon; it forms a sheer, vertical face sometimes up to 76 meters (250 feet) high along the lower walls.

The characteristics of this geological layering affect availability and distribution of water—a critical resource in this arid land and an important factor for human occupation. The highly porous Navajo sandstone serves as an

The sheer red cliffs of the Tsegi Canyon system are Navajo sandstone. The Kayenta formation forms the bench just below.

aquifer for collecting and storing water. Rainwater percolates down through this permeable rock until it reaches beds of mud and shale in the Kayenta formation below. Unable to pass through this impervious layer, the water flows out horizontally, coming into the canyons as seeps or springs. Sometimes canyon bottom alluvium traps the water, forming a high water table ideal for supporting a lush growth of natural vegetation or providing a dependable water supply for agriculture.

The high plateau country receives about half of its annual precipitation in the form of winter snows.

The "pinyon-juniper pygmy forest" is the predominant vegetation in high plateau country of the northern Navajo reservation — at about 1500 to 2100 meters (5000 to 7000 feet) above sea level. All plants in this pygmy forest

Erosion by wind and water reveals the lines of ancient wind-blown sand dunes in the Navajo sandstone.

have to be adapted to a semi-arid environment, as the area only averages about 250 to 300 millimeters (10 to 12 inches) of rainfall per year. Vegetation is stunted and sparse. Pinyon pine and juniper (commonly called cedar) are the dominant plants in this community; they rarely reach a height of more than 9 to 12 meters (30 to 40 feet). Interspersed with these pygmy conifers are

desert shrubs such as big sagebrush and a variety of grasses. The pinyon-juniper pygmy forest covers the high mesas through which the Tsegi and Navajo canyons cut.

The lushness of Betatakin Canyon, where Betatakin ruin is located, contrasts sharply with mesa top vegetation. Most striking is the large stand of quaking aspen lining the canyon bottom; the white trunks and leafy greenery stand out dramatically against the red walls of the canyon. The pinyon and juniper of the mesa give way gradually down the slope to the Gambel's oak and box-elder which, along with aspen, predominate along the bottom. Large Douglas-firs grow on the north-facing slope. The intermittent wash is choked with

A gnarled juniper tree — part of the pinyon-juniper pygmy forest that covers the mesas.

"Today the harsh beauty of the steep cliffs and canyons strikes the visitor; but this was of little practical value to the Anasazi, trying to make a living in an inhospitable land."

Tall aspen trees in Betatakin Canyon offer a pleasant surprise for visitors to the arid canyon country, where sparse and stunted vegetation usually predominates.

Hopi symbol—sunflower in circle of life symbolizing youth.

streamside plants such as horsetail, willow, and meadow rue.

The high water table in the canyon bottom supports this almost jungle-like vegetation. The fact that Betatakin is a narrow canyon is also important; the steep walls block out much of the day's sunlight. As a result, there is less moisture loss by evaporation, and more moisture and humidity present to maintain the abundant plant life. Vegetation in Betatakin Canyon is much the same as it was 700 years ago when the Anasazi were living there.

The rest of the Tsegi canyon system, however, including Keet Seel Canyon where Keet Seel ruin is, has undergone severe erosion in the past hundred years, changing its aspect dramatically. At the turn of the last century the canyons were much greener, with several small ponds inhabited by ducks and other waterfowl located in the lower reaches. A combination of recurring drought conditions and heavy grazing by Navajo livestock has caused the recent erosion. Because of the resulting drop in the water table, the bench above the arroyo can now support only a sparse cover of sagebrush and a few grasses.

The prehistoric environment of the Tsegi and in particular of Keet Seel Canyon was also lusher than that of today. The Anasazi used aspen as a building material at Keet Seel, which indicates it was growing in the immediate

Because erosion has caused the water table to drop in Keet Seel Canyon, aspen can no longer grow in the area around Keet Seel.

area. Today small stands of Gambel's oak and a few boxelder grow near the ruin, but aspen are totally absent because there is not enough moisture to support them. Evidence shows that there was also a cycle of erosion in prehistoric times which culminated around A.D. 1300, just about when the Anasazi left. The alluvium was redeposited in the 500 years between that date and the beginning of the present erosion cycle.

Nitsin Canyon, the branch of the Navajo Canyon system where Inscription House ruin is located, is about 300 meters (1000 feet) lower in elevation than the Tsegi. Overgrazing by livestock has also affected Navajo Canyon and its tributaries. Today the barren flats below Inscription House are cut by deep arroyos and thinly covered with sagebrush and Russian thistle. The prehistoric environment of Nitsin Canyon is still unknown, as no research has been done on it to date.

Today the harsh beauty of the steep cliffs and deep canyons strikes the visitor; but this had little practical value for the Anasazi, trying to make a living in an inhospitable land. Rainfall is scant and undependable. When it does come, it is often in violent thundershowers that can wash out carefully tended fields. At 2100 meters (7000 feet) above sea level, the range of temperatures is extreme; cold, snowy winters alternate with hot, dry summers. In some years early and late frosts do severe damage to crops and natural vegetation. Facing these natural hardships, the Anasazi had only their wits and perseverance to insure their survival.

ANASAZI

"Some experts believe that influences from the highly developed cultures of Mexico and Central America provided the impetus for many developments in the prehistoric Southwest."

The Anasazi occupied what is now called the Four Corners area — the high plateau country near the Colorado and San Juan Rivers where today the states of Arizona, Colorado, New Mexico, and Utah meet. Where did they come from? They probably descended from the Paleo-Indians who roamed the West at least 10,000 years ago. These ancient, pre-Anasazi hunters used large, simple stone tools and stalked big game animals like bison, camel, and mammoth.

As most of the larger game animals became extinct, early Indians turned to greater use of wild plant foods, while continuing to hunt rabbits, birds, deer, mountain sheep, and other available game. Archeologists refer to this hunting and gathering way of life that developed as the Desert Culture. It was probably the base out of which the Anasazi tradition grew.

Archeologists have divided Anasazi culture into two periods — Basketmaker and Pueblo. It is difficult to assign exact dates to the various stages of development, since different groups of these people changed at differing rates and at different times. The first remains identifiable as Anasazi date to about A.D. 1, and the Basketmaker period lasted until about A.D. 700. Pueblo culture reached its peak in the 1100's and 1200's — the so-called Classic Period. The Anasazi abandoned the Colorado–San Juan River area around A.D. 1300, but this did not mark the end of the Pueblo people. Pueblo cultures, continuations of the ancient Anasazi tradition, still carry on today in the Hopi and Zuni areas to the south and along the Rio Grande in New Mexico.

Anasazi culture did not exist in isolation. Farther south — in what is now southern Arizona and New Mexico — the Hohokam and Mogollon traditions also developed out of a Desert Culture base. Similarities and parallels among these three major Southwestern traditions suggests there was considerable contact between them.

Some experts believe that influences from the highly evolved cultures of Mexico and Central America provided the impetus for many developments in the prehistoric Southwest. Most agree that, at the least, agriculture and pottery-making probably originated among these southern peoples and spread north — first to the Hohokam and the Mogollon, then to the more distant Anasazi. Travel, trade, and possibly small-scale migrations were the means for exchange of both goods and ideas. By the time these traits traveled from Mexico to the Anasazi of Tsegi Canyon, they had passed through many hands and changed considerably in the process — sometimes to the point of obscuring their original source.

Basketmakers

Excavation of overhanging recesses in

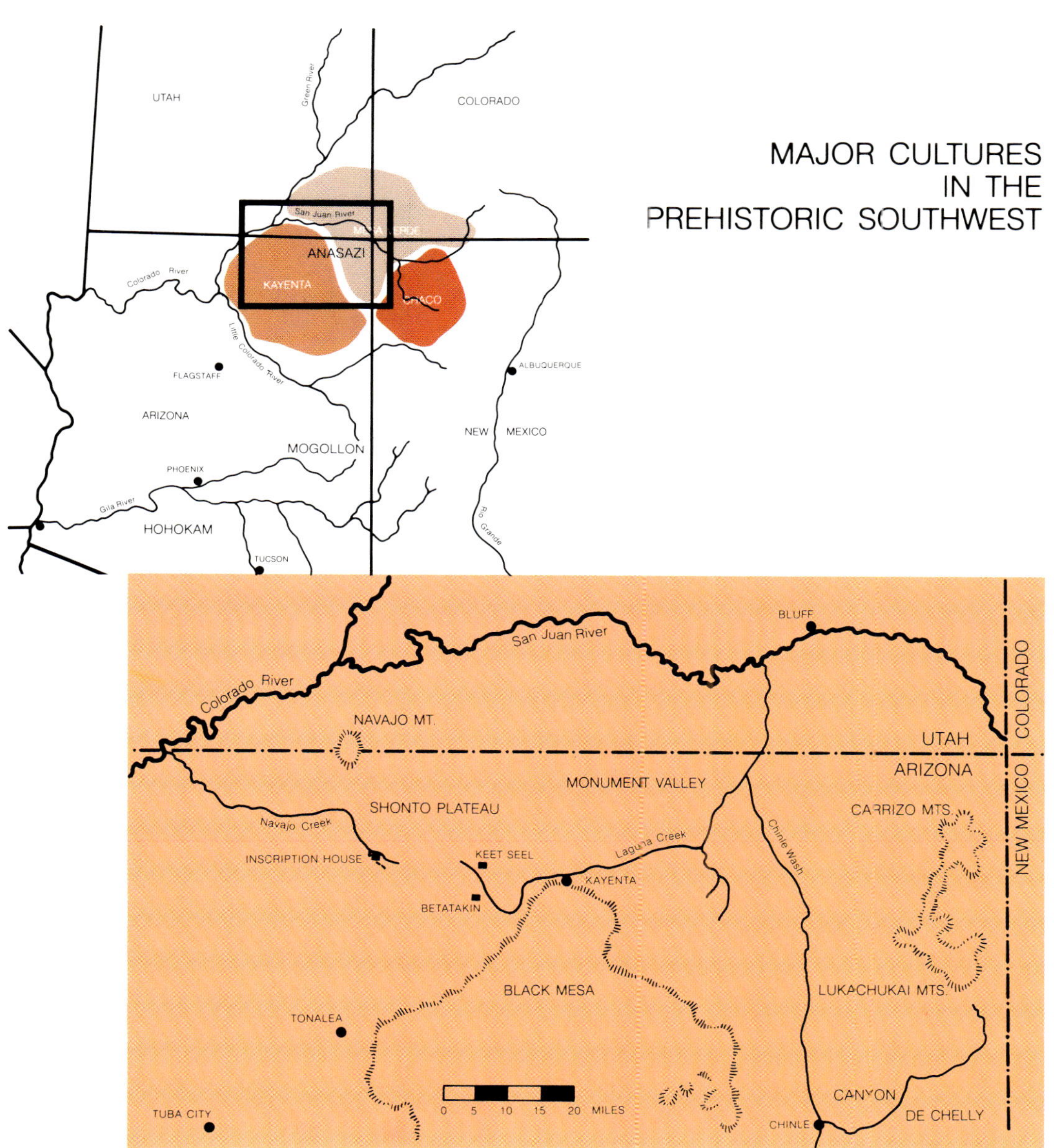

MAJOR CULTURES
IN THE
PREHISTORIC SOUTHWEST

"In order to make a living, a hunting and gathering people like the early Basketmakers had to have an intimate familiarity with the land—its flora and fauna and its cycles—that would astound most modern people."

the sandstone cliffs turned up the earliest evidence of the Anasazi. It revealed a people who lived primarily off the land — hunters and gatherers. In contrast to Desert Culture groups, however, these people had begun to cultivate certain plants to supplement their diet of wild foods. They made elaborate and fine baskets, and because of this are referred to as the Basketmakers. Archeologists have found some of the clearest evidence of Basketmaker culture in Monument Valley east of Tsegi Canyon.

The plateau country offered abundant plant and animal food for a hunting and gathering people like the early Basketmakers. But to make a living in this way required an intimate familiarity with the land — its flora and fauna and its cycles — that would astound most modern people. Basketmaker people had to know just when and where the various plant foods were ready to harvest. They undoubtedly spent many long, hard hours collecting enough roots, berries, acorns, seeds, pinyon nuts, and cactus fruits to provide an adequate meal for the family.

Basketmakers hunted the ever-abundant rabbit with nets, snares, and throwing sticks. For stalking larger game such as deer and mountain sheep they used the atlatl or spear-thrower. The hunter held this simple device in his hand to extend the throwing length of his arm, allowing him to propel a spear farther and more accurately. Though better than the unaided arm, the atlatl was less efficient than the bow and arrow, which did not become common until later Anasazi time. Even as they became more dependent on farming for their livelihood, the Anasazi people continued to supplement their diet with wild foods and to fall back on hunting and gathering when their harvest was poor.

Though their sharp spines may seem unfriendly, local cactuses produce beautiful flowers in the spring and many yield tasty fruits in the fall.

They learned about agriculture at an early date but it was by slow degrees, over a period of hundreds of years, that it gradually became a central part of their livelihood. At first, the Basketmaker people planted fields of corn and squash which they often tended only sporadically, while continuing to depend primarily on hunting and gathering. With time, however, they devoted more and more energy to farming, and it began to account for a larger part of their food supply. They added variety to their diet when they started to grow beans in addition to corn and squash.

The earliest Basketmakers spent a great deal of time on the move in search of wild plant and animal foods. Because of this, they traveled in small groups and stayed in cliff overhangs or temporary shelters. As they began to settle down more to take care of their fields, the later Basketmakers built partially underground houses — or "pithouses." Some features of the pithouse appear to have a long history that stretches back thousands of years, across the North American continent and into parts of eastern Asia. The builder covered the dwelling over with sticks and logs to form a roof, and often lined the pit with sandstone slabs. Late Basketmakers lived a more sedentary lifestyle in relatively permanent villages of several such pithouses.

At first the Basketmakers made no pottery, using their baskets for daily tasks like storage and cooking. They wove baskets in an assortment of different shapes — shallow trays, jars, bowls, bags, and others. Weavers used sturdy but pliable materials like willow twigs and hardy grasses.

They made some of their finest baskets by a coiling technique, building them up from the bottom in spiraling coils which they sewed together with thin splints. Some of these were woven so tightly they could hold water. Often the weaver applied pine pitch to make the basket even more watertight. Since baskets could not be placed directly on the fire for cooking, the Basketmaker woman took hot rocks from the hearth and dropped them into a basket full of water and food. The stones heated the water and cooked the food.

They also made other serviceable items from natural fibers. One of the most useful plants was the yucca. The Indians scraped or chewed on the leaves to separate the strong threads, and then twisted the strands into strings and ropes. They wrapped some of these strings with rabbit fur and wove them into blankets to keep warm during the cold winter months. The Basketmakers are noted for the excellence of the square-toed sandals they fashioned from yucca fibers. Footwear was crucial for protection in the desert, but their sandals show a degree of craftsmanship that goes beyond the purely functional.

Most of what we know about the

"Most Anasazi people would probably see little change in their entire lifetime. Life would have seemed much the same from day to day, and from year to year."

The Kayenta Anasazi continued to build pithouses up into late Pueblo times — often right next to the above-ground pueblo structures. This partially reconstructed pithouse at Mesa Verde National Park dates to Basketmaker times.

Basketmakers and their material goods is the result of burial practices. They placed their dead in the pits dug for storage in the floor of the cliff overhangs; here the burials were well protected from moisture. Because of this, many "mummies" have lasted to the present in a remarkable state of preservation, even though Basketmakers did not embalm or otherwise artificially treat their dead.

Bodies were tightly flexed, with knees to chest. They often wrapped the dead person in a fur blanket, placed a large basket over his head, and set a pair of new sandals next to him. Though we know nothing about Basketmaker beliefs concerning the supernatural, the frequency of these and more elaborate burial offerings speaks of some faith in an afterlife.

These examples of Basketmaker artifacts show the fine workmanship of the early Anasazi. The atlatl—or throwing stick, though warped, is otherwise well preserved.

Today we take change for granted. The world around us and often our personal lives, too, are changing all the time and at a dizzying speed. Or so it would seem to an Anasazi. For him, life would have seemed much the same from day to day, and from year to year. Most Anasazi people would probably see little change in an entire lifetime. But things do change and cultures do evolve, even though slowly, for no way of life is totally static. Sometime around the 600's, new influences came into the Southwest—probably from Mexico—that greatly altered the lifestyle of the Anasazi people.

Pueblos

Anasazi culture was not exactly the same from area to area, for people naturally responded to local conditions and influences. It was during the Pueblo period that distinct regional variations began to develop. These were more differences of style than of substance. Anasazi development was remarkable because it did show considerable uniformity over a large geographic area.

However, based primarily on variations in architecture, pottery, and geographic location, archeologists have distinguished three major Anasazi areas. The Mesa Verde area is represented at Mesa Verde National Park in Southwestern Colorado, and the Chaco area at Chaco Canyon National Monument in northwestern New Mexico. Betatakin, Keet Seel, and Inscription House at Navajo National Monument in north-

Settlement, growth and final abandonment of Betatakin all took place in less than two generations.

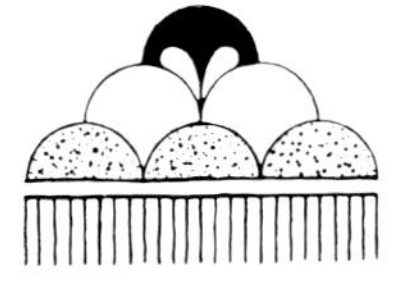

Hopi symbol — clouds and rain with kachina overseeing.

eastern Arizona are examples of developments in the Kayenta area.

The environment in the Kayenta area was rigorous, and making a living was not easy for its people. They were a long way from many of the established routes of travel and trade — especially those with Mexico. These factors may have kept some aspects of Kayenta culture on a simpler level than that of the other Anasazi branches. Kayenta society, for instance, was probably less complex than that of Chaco, if only because people were fewer and less concentrated in the Kayenta area. In other things, however — such as pottery-making — Kayenta people equalled or excelled their Anasazi neighbors. It might be fair to say that Kayenta culture was in some ways less sophisticated than the other branches — especially the Chacoan; but to call the Kayenta people backward would be a mistake.

The gradual evolution of a more settled, agricultural way of life during the Basketmaker period had set the stage for the growth of villages and towns in the Pueblo period. Farming calls for a more sedentary lifestyle than hunting and gathering. It can also support more people with a more dependable food supply, so they began gathering together into larger and more permanent villages.

Settlements of the early Pueblo period, however, were still small and scattered — not a lot different from late Basketmaker villages. One of the most noticeable differences would have been the appearance of more buildings above ground. The Anasazi probably first built surface rooms for storage, but soon established living quarters in them as well. Later they constructed rows of adjoining rooms. One or two such rows of 12 to 14 rooms and maybe a few semisubterranean structures would make up an entire village.

Gradually villages became fewer in number and larger in size. Most ranged between 10 and 40 rooms, but some had up to 200 rooms. Growth of towns in the Kayenta region never reached the extremes of size and urbanization found in the Chaco area. The land probably could not support such large concentrations of people. Settlements remained smaller and their population numbers more modest.

Why did the builders of cliff dwellings like Betatakin, Keet Seel, and Inscription House locate their towns in such high and relatively inaccessible spots? One explanation proposed is that these sites were defensive. There is no evidence, however, of enemy people in the area at the time, though jealousy and strife between neighboring villages is a possibility. Other than their elevated position, the cliff villages in Tsegi Canyon show no indications of provision for defense.

On the other hand, there are ample good reasons why the people might

have chosen to build in the cliff shelters. Farm land was at a premium in the narrow canyons, and cliff sites were conveniently close to, but not right on, arable land. The overhangs provided shelter from the elements. Walls and belongings were protected from damaging winds and rains, and the people could carry on their daily activities in the open village courtyards regardless of the weather. The water that seeps down through the Navajo sandstone, helping to form the alcoves in the cliffs, emerges as springs at the base of the formation. These springs associated with cliff shelters were another attraction for the Anasazi.

The particular alcoves in which they chose to locate almost invariably face south or southeast. In the heat of summer when the sun is high, these sites are in the shade. During colder months, the low winter sun shines directly into the overhang. In a land of climatic extremes, these strategically chosen locations in cliff shelters provided a natural temperature control.

Building techniques of the Kayenta Anasazi are one of the important characteristics distinguishing them from other Anasazi branches. The Chacoans were the finest Anasazi masons; their walls show attention to excellence and appearance that far surpasses other Anasazi efforts. Kayenta masonry, however, is generally poorer than that of either of the other Anasazi branches.

Salvage archeology along the monument entrance road right-of-way uncovered this early Pueblo site. Rows of rocks show where the stone walls of the structure once stood. An extended family group probably occupied the small pueblo.

"The gradual evolution of a more settled, agricultural way of life during the Basketmaker period had set the stage for the growth of villages and towns in the Pueblo period."

Their architecture is adequate but not elaborate.

Kayenta masons did not bother to shape building stones carefully or uniformly, and they applied the mud mortar liberally. Walls were not always perfectly straight or the corners quite square. Builders used a technique known as "chinking"—setting small rocks or even potsherds into the binding mortar—to provide a stronger, sturdier wall.

Kayenta Anasazi builders did not devote the same time and care to constructing masonry walls as their Anasazi neighbors in the Chaco Canyon area did.

The first Anasazi walls were of wattle and daub construction. To make these they set a row of upright sticks, which they bound together with twigs or string and covered with mud. Though not as sturdy or long-lasting as stone masonry, this simpler type of wall continues to appear even in late Kayenta cliff dwellings. This seems to be further evidence of the Kayenta peoples' apparent lack of concern with the quality of their architecture.

Living rooms can be identified by smoke blackening on the inside walls. Doors were small for better insulation against the heat and cold, and smoke holes were small or nonexistent. As a result, ventilation was poor. Soot from indoor fires probably coated lungs as well as walls. "Bedrooms" might be a better term for these rooms, as the Anasazi generally used them only for sleeping. Most are uncomfortably small, and not pleasant to spend much time in. Day-to-day work was done in open areas of the village.

The Pueblo people kept their food in well-built granaries, taking care to protect their stores from rodent and insect pests. Builders elevated door sills several feet above ground level. Sandstone slabs fitted into grooves around the door sealed off the room. Most granaries appear to belong to individual families, but some village sites seem to have had a communal granary

used by the whole group.

The Anasazi of Tsegi Canyon grew their crops — corn, squash, several varieties of beans, and a few native plants like beeweed — on the broad alluvial flats along canyon bottoms. Their agricultural technology was simple. The farmer planted seeds with a digging stick. The high water table at the bottom of the canyon made dry farming feasible most of the time.

The mud has worn off the upper part of this wattle and daub wall to reveal the stick construction beneath.

Whereas today the creek flows at the bottom of a deep gully as the result of severe erosion, then it meandered across an uninterrupted surface that stretched from one canyon wall to the other. There is no material evidence of irrigation in the Tsegi, but Anasazi farmers could have easily redirected stream water to their fields by means of ditches. Techniques of water diversion were familiar to many of their close neighbors, and it is likely the Tsegi people practiced them, too.

Corn was the staple of Anasazi diet. They roasted some of it and ate it right off the cob. They also made corn into meal for flat cakes and mush. Women ground the kernels between a large slab of rock called a "metate" and a hand-held stone called a "mano." Sometimes metates were set in a series of bins, with surfaces graded from rough to smooth to produce a progressively finer meal. Like Hopi women in historic times, an Anasazi woman undoubtedly spent many hours kneeling

The Anasazi sealed off these granary doors at Keet Seel with fitted sandstone slabs to protect their food stores.

This roof at Keet Seel shows how the Anasazi used beams, sticks, juniper bark matting, and mud to cover their rooms.

by these bins to prepare dinner for her family. The women of a household may have gathered in a common family courtyard or grinding room, to visit and exchange gossip as they worked.

The prehistoric Anasazi knew nothing of sheep, cattle, or horses, as these were not introduced to the Southwest until the Spaniards arrived in the 1500's. They did have tame dogs, as pets and possibly for hunting. Domesticated turkeys provided feathers for blankets, ceremonies, and personal adornment. No one knows whether the Kayenta Anasazi ate these birds.

The Pueblo people continued to make fine baskets; but more and more, pottery began to replace woven containers for many daily uses. For the early Basketmakers, who moved around a lot, baskets were more practical than heavy and easily broken pots. However, the advantages of pottery for tasks like cooking and storage quickly became apparent to the Anasazi as their lifestyle became more sedentary.

They never developed a potter's wheel, making pottery instead by a coil and scrape technique. The potter — usually a woman — rolled the clay into long "ropes" which she then set in coils, building up in layers to form the shape that she wanted. To finish the surface, she scraped the coils smooth.

In middle Pueblo times, potters began to leave some or all of the coils showing for decoration on the outside of the pot. Sometimes they pinched or incised simple patterns in the coils. This corrugated ware and the more basic plain gray ware served as the functional, everyday ware in which the women cooked.

It was in the ceramic arts that the Kayenta people excelled. Painted black-and-white pottery was typical of all the Anasazi, but in the hands of Kayenta potters it reached a quality unparalleled elsewhere. Their geometric designs show imagination and skill. Some of the later pots have so much black design that they look more like white-on-black.

It was not with the black-on-white pottery, however, that Kayenta craftsmanship reached its peak. There is a black-on-red ware found in the Kayenta area which archeologists believe originated in the south. In late Pueblo times, the Kayenta potters combined this tradition with the black-on-white to produce a polychrome — or "many-colored" — ware of black, red, and sometimes white on orange. This polychrome, unique to the area, is remarkable for the elaborateness and intricacy of its design. Many people consider it the finest, most beautiful Anasazi ware made.

"Rock art" probably provided another medium for Anasazi artists. Though they had no formal written language, they left signs and pictures on the canyon walls. Some — called pictographs — they painted on the surface. Others — called petroglyphs — they incised into the rock. We do not fully understand the function of this rock art. Could it be a message left behind for other people, or — like pottery decoration — a form of artistic expression? Perhaps some representations involved rituals or magic.

Whatever it meant to the Anasazi who made it, rock art is a source of fascination to people today. Even more than ancient walls and pottery, it seems to strike a common chord and to speak of the universal and timeless urge of man to communicate with his fellows.

What did the people themselves look like? The Anasazi of the Pueblo period wore little clothing most of the time. The usual attire for men was probably just a breechclout, and for women a sort of apron. They fashioned these of woven fibers, cotton cloth, or deerskin. The Pueblo people still made sandals

Hopi symbol — circle around butterfly symbolic of youth.

Historic Hopi use the same simple agricultural methods that their Anasazi ancestors probably did. They situate their small farming plots carefully to catch runoff from scanty summer rainfall.

to protect their feet. They also wove blankets of cordage, wrapped with feathers as well as fur. People enjoyed adorning themselves with jewelry — including bracelets, beads, and pendants of bone, shell, and turquoise.

No one knows whether the Anasazi of Tsegi Canyon grew cotton themselves or traded for it from farther south; but they did weave fine cloth from its fibers. Among the Hopi, the men traditionally do the weaving in the ceremonial rooms. Archeologists assume the practice was the same among the Anasazi. They have found large cotton items such as ponchos and blankets at some Kayenta Anasazi sites.

Researchers have put together a sketchy picture of what Pueblo people were like physically. They were several inches shorter than the average person today. Many suffered from ailments of the bones and joints, such as arthritis. They lost a third of their children before the age of five. Average life expectancy was only about 45 years. Life was not easy for these people, and poor nutrition, disease, and accidents took their toll.

At some point, they took to tying their small children tightly into hard cradleboards. This practice flattened the back of the skull as it grew, giving the head a broader appearance. The Anasazi seem to have done this intentionally. Perhaps they considered it attractive, and it was fashionable at the time.

Skeletal remains show that they suffered from terrible dental problems. Sand in the grinding bins accounted for much of the trouble. Every meal included grit that had mixed with corn in the grinding process. Year after year of chewing on this hard sand wore away teeth and undoubtedly caused a great deal of pain.

Anasazi women ground corn into meal in these bins, using metates and manos. Imagine how much sand mixed into the corn meal in the process!

One mystery of the Anasazi is that, in many areas, archeologists have not found nearly enough burials to match population estimates. This is particularly true in the Tsegi; excavation has turned up only a few burials at both Betatakin and Keet Seel. Archeologists have not made a thorough search of the canyons, so it is possible other burials may be located away from the immediate village areas. There is no evidence that the Kayenta Anasazi practiced cremation.

A number of the burials have been uncovered in trash mounds. This common Anasazi practice was not necessarily a token of disrespect. The trash mound was probably the easiest place to dig, and the people may not have regarded it with the same distaste that we do. Could they have considered it a sacred area where they disposed of things that were once good and useful?

When the Anasazi gathered together into compact villages their society grew more complex. Necessity for interaction and cooperation among the people of a town caused many of the changes. More elaborate social ties and expectations evolved and spread. These helped people get together on projects that benefited the whole village, and also cut down on competition and conflict.

The rooms at Betatakin and Keet Seel are grouped into clusters — usually one or more living rooms and several storage rooms, all opening onto a central courtyard or work area. Each cluster was probably home to a household group. Most of our households today include just one set of parents and children, but among the Anasazi they may also have included the wife's parents plus her sisters and their families.

"The impersonal nature of today's mass society would be unimaginable to an Anasazi; but the emphasis on freedom and individuality in our society would seem equally strange."

Like the Hopi, the Kayenta Anasazi probably traced decent through the women of the family; family belongings passed from mother to daughter, rather than from father to son as is traditional in our society. When a couple married, they usually went to live with the wife's family.

These household groups may have been linked together into a larger kinship group, such as a lineage — or family line — related through their womenfolk. Several such lineages could make up a clan, named after an ancient ancestor — like the Badger Clan or the Flute Clan among the Hopi. Members of a clan all consider themselves related, even though they may not be able to trace biological ties.

In addition to his responsibilities to his relatives, a man also had obligations to help out the other men of his religious group with ceremonies. These cross-cutting ties helped to hold the pueblo together.

Anasazi towns were undoubedly very close-knit communities, where everyone knew what everyone else was up to. The society seems to have been egalitarian — no one person had a lot more power or wealth than the next. The pressure to conform and live up to group ideals was probably strong. Religious beliefs, family pressures, and village gossip served effectively to keep most people in line. The impersonal nature of today's mass society would be unimaginable to an Anasazi; but the emphasis on freedom and individuality in our society would seem equally strange.

The Hopi today hold most of their sacred rites in subterranean ceremonial chambers call "kivas." These specialized rooms also serve as clubhouses for men of the village. Kivas have an ancient tradition, for they also occur in prehistoric Anasazi villages. Those built by the Kayenta people

A household group probably occupied this room cluster at the east end of Betatakin. Smoke blackening on interior walls identifies the closer structure as a living room. Rooms with roof entrances are storage chambers, not kivas.

tended to be simpler than their counterparts in other Anasazi areas. Certainly Kayenta kivas were not as numerous as kivas elsewhere. Were the Kayenta people perhaps less concerned with religion than other Anasazi were?

A visitor to Mesa Verde or Chaco Canyon is likely to come away with the feeling that all kivas are round. But in the Kayenta area kivas come in a variety of shapes. Rectangular kivas were common, especially in the later sites like Betatakin and Keet Seel. Kivas in Hopi villages today are also rectangular.

Certain features distinguish kivas from other rooms in an Anasazi village, though among Kayenta people these features were less consistent than elsewhere. To get into the kiva, they climbed down a ladder through a roof hatchway. Kiva ceilings were high enough to allow a person to stand erect. At one end of the room was a ventilator shaft to let in fresh air. An upright stone slab between the vent and the fireplace served as a deflector to keep drafts from scattering embers and ash from the hearth. Usually a bench encircled all or part of the kiva. Storage niches in the walls held ceremonial objects like beads, clay figurines, and prayer feathers. Kiva floor features sometimes included a set of loom anchors.

What the Hopi call a "sipapu" is a single hole in the kiva floor. The sipapu is the place of emergence—the opening through which the Hopi ancestors came up from the world below. The original is just off the Colorado River in the Grand Canyon, and the sipapu in the kiva is a symbolic representation of it. It also serves as a path for communication and travel between the kiva and the sacred underworld.

Many stylized kiva features can be

The Hopi—like the Anasazi—use ladders to climb down into their kivas. These kivas are on the south end of the Hopi village of Mishongnovi. The picture was taken in the early 1900's.

traced back to the pithouse, which was once the center of both domestic and ceremonial life for the Anasazi. Some people apparently continued to hold religious ceremonies in the underground rooms, even after they had moved into surface dwellings. Once specialized ceremonial rooms—or kivas—evolved, they spread throughout the Anasazi area and became a common feature of all Pueblo villages.

Farming left the Anasazi more free time, especially in winter months. Also, there were more people in the towns to participate in ceremonies. As a result, religion flourished in their pueblos. Prehistoric religions are difficult to learn about in any detail, as they leave little material evidence behind. Kivas and ceremonial objects are suggestive, but they do not tell us about the stories and ideas of the people. We can imagine, though, that the elaborate legends and ceremonialism of the historic Hopi have roots in the beliefs and practices of these earlier Anasazi town dwellers. No doubt an awareness of the supernatural was an important part of their daily lives, as it is today for the traditional Hopi.

Hopi symbol—eagle, used only on pottery.

Hopi religion is highly formalized. Its focus is on the yearly cycle of ceremonials that maintain the balance of the universe and bring rain to make their crops grow. Anasazi farmers probably held similar kinds of rituals in their kivas, though perhaps on a simpler level. Archeologists have found no evidence of the colorful and elaborate Pueblo Kachina cult among these Hopi ancestors. They think it was introduced—possibly from Mexico—sometime after the Hopi settled in their present homeland.

All in all, the Pueblo Anasazi seem to have been leading a good life in the Tsegi Canyon and over the Colorado Plateau. They were enjoying the benefits of new levels of culture and technological improvement. Life was still not easy, but certainly better. Their achievements are a lesson for people who still think the New World that Columbus discovered in 1492 was populated by "primitive savages."

So what happened to make them leave?

This is a question Southwestern archeologists have puzzled over for years. The Anasazi left the Kayenta region around A.D. 1300. By this time, Anasazi peoples had moved out of the Mesa Verde and Chaco areas, too, leaving the entire San Juan–Colorado River area deserted. Researchers still do not know the full story behind this abandonment, and probably never will. It is likely there was no one simple cause, but rather a number of contributing factors.

Archeologists have come up with many theories. Some think that nomadic, warlike Indians came in and preyed on the peaceful Anasazi farmers, finally driving them out. They have

not been able to back this up, however, with concrete evidence that such people were in the area at the time. Tree-ring evidence shows there was a bad drought during the last part of the 13th century. An unfavorable shift in the seasonal rainfall pattern also seems to have occurred at about the same time. Could these factors have caused a severe famine and forced the Anasazi to move in search of more water and better conditions? The lands which they farmed were already marginal, and overuse may have depleted the soil and other resources. We also should not underestimate the power of social factors and religious beliefs to influence the actions of people, even though these things are intangible and leave no material evidence.

Summer thundershowers cause flash floods, which sometimes tumble over canyon walls to form waterfalls. Rushing water is the major cause of erosion in the canyons.

In the Tsegi Canyon, researchers have found indications of severe erosion at about the time that tree-ring evidence shows the Anasazi moved away. Some archeologists think this erosion was an immediate cause for the departure. Drought conditions probably reduced the plant cover. In addition, the Pueblo people cleared large areas of vegetation to plant their crops. They cut down many trees for building and firewood, even to the point of apparently denuding sections of the canyon. When heavy summer thundershowers came there was little vegetation to catch running water from torrential rains and hold the soil in place. Rains carried the earth downstream, and over the years cut deep arroyos in the alluvial flats.

This erosion was no doubt disastrous for Anasazi farming. The ground water on which they depended to support their crops probably dropped severely. With the stream running at the bottom of the arroyo, farmers could not divert water for irrigation. In addition, gullies may have gouged through the middle of their fields, further reducing already scarce farmland in the narrow canyons.

"Rather than radically change their way of life, perhaps the Tsegi people preferred to move to the Hopi mesa area, where the environment was more stable."

This may have presented the Anasazi with a serious dilemma. Their technology had promoted the growth of larger towns and a more complex society. But now their farming techniques were no longer adapted to the changing environment. To support even part of the population in the Tsegi area, villages like Betatakin and Keet Seel would have to break up into smaller groups. Their society would have to reorganize on a simpler level.

Under more favorable circumstances, social relationships had gradually developed which held the Anasazi towns together. People were probably unhappy with the prospect of disrupting these ties. If the threat of famine forced some to leave in search of a better life elsewhere, relatives and friends may have followed in order to be with them. Rather than radically change their way of life, perhaps the Tsegi people preferred to move to the Hopi mesa area, where the environment was more stable and their farming techniques could still support the village organization they were accustomed to.

How much did the Anasazi themselves have to do with environmental changes that occurred in the Tsegi? We have recently begun to look around us and to see how man's exploitation of the earth's resources has upset the balance of nature. It is natural to try to apply our new insights to the Anasazi and their situation. Certainly the concentration of populations in villages like Betatakin and Keet Seel intensified the impact in those areas.

Anasazi technology was simple, but it did have some effect on the land. Practices such as land clearing were not the only cause of erosion in Tsegi Canyon; but they may have contributed to and hastened development of the problem. To place the blame on the Anasazi for their own departure, however, is perhaps to overemphasize their role in relation to other factors such as climate and topography.

CLIFF VILLAGES

Cliff villages like Betatakin blend fully with their surroundings, for the Anasazi constructed them of the same red rock in which they were set.

Betatakin

Anasazi people must have found the huge sandstone alcove in Betatakin Canyon an excellent village location. Certainly the sheltering overhang and a flowing spring were attractions. Just a mile away, the canyon mouth opens into the main Tsegi Canyon, where the flat, open land and running stream were ideal for growing crops. Here was a good spot to farm and raise a family.

The first occupants of the Betatakin rock shelter arrived sometime in the 1250's.They were few in number and left little imprint. Perhaps they just stayed seasonally to grow crops in the Tsegi in order to help feed families living in their distant village. They may have told other people in their home community about the fine location that they found; for between A.D. 1267 and 1268, three or four households settled in the alcove.

"The Betatakin village was a tight, homogeneous community. The move to the alcove appears to have been a planned and deliberate one by a single group of people."

These pioneer families seem to have been a vanguard, sent ahead to prepare the rock shelter for occupation. They cut and stockpiled extra beams for use in future construction, and probably cleared many years' accumulation of debris. Seven years later, a major immigration began that brought the population of Betatakin to 75 to 100 people by A.D. 1277. These people, presumably from the same village as the first families, used the stockpiled beams to build their new homes.

The move to Betatakin alcove appears to have been a planned and deliberate one. The Anasazi who settled here probably came from some other established community elsewhere. We donot know why these people decided to leave their previous homes — perhaps because of farmland erosion or some sort of social pressures. Whatever the reason, it was not too pressing; for they took the time to carefully choose and prepare their new village site.

The social ties that held the old village together must have been strong. In the face of adverse conditions, they chose to relocate the group as a whole rather than splinter into individual family units. In addition, they had a way to make and carry out such an important group decision. This probably required some kind of strong leadership.

Building techniques at Betatakin are consistent. This is further evidence that the Anasazi who settled the village were all members of the same community. Only people who had lived together before would be likely to share such uniform architectural standards.

After the major influx of people, Betatakin grew more slowly to reach a peak of about 125 inhabitants. The arrival of occasional immigrants caused some of this later growth, but natural population increase was probably more important. Children grew up and babies were born. No doubt newly-wed couples setting up their own households built many of the new room clusters.

Betatakin at its height was a bustling community. Today most of the

Though the Betatakin gallery might appear at first glance to be a lookout, in fact, it commands a limited view. More likely the Anasazi stored food behind its walls.

rooms in the east-central part of the village are gone — swept away by a rockfall; but in the late 13th century, over 100 rooms sprawled in an arc along the steep floor of the rock shelter and filled every available space. A few late rooms even stretch beyond the alcove to the east, outside the shelter of the overhang. Though now the village seems unnaturally silent, any visitor who has tried an experimental echo in the shelter can appreciate how the daily activities of 125 Anasazi men, women, and children (not to mention dogs and turkeys!) filled it with noise and commotion.

Considering this relatively large population, it is surprising that archeologists have positively identified only one kiva at Betatakin. A smaller kiva in an alcove up the canyon from the main village also served the people of the town. Both kivas are rectangular.

We do not know exactly when the Anasazi built the kivas. The first residents of Betatakin may have traveled back to their old village to participate in the ceremonies there. The people had probably built at least one kiva of their own by A.D. 1277, when the main immi-

Rooms at the far western end of the Betatakin alcove.

igration was completed. Perhaps they added the second ceremonial chamber

"Though now the village seems unnaturally silent, any visitor who has tried an experimental echo in the shelter can appreciate how the daily activities of 125 Anasazi men, women and children (not to mention dogs and turkeys!) filled it with noise and commotion."

When the Anasazi lived at Betatakin, rooms occupied the whole alcove. Sometime after they left, a large sandstone slab broke loose from the rear of the rock shelter and wiped out a large section of the village.

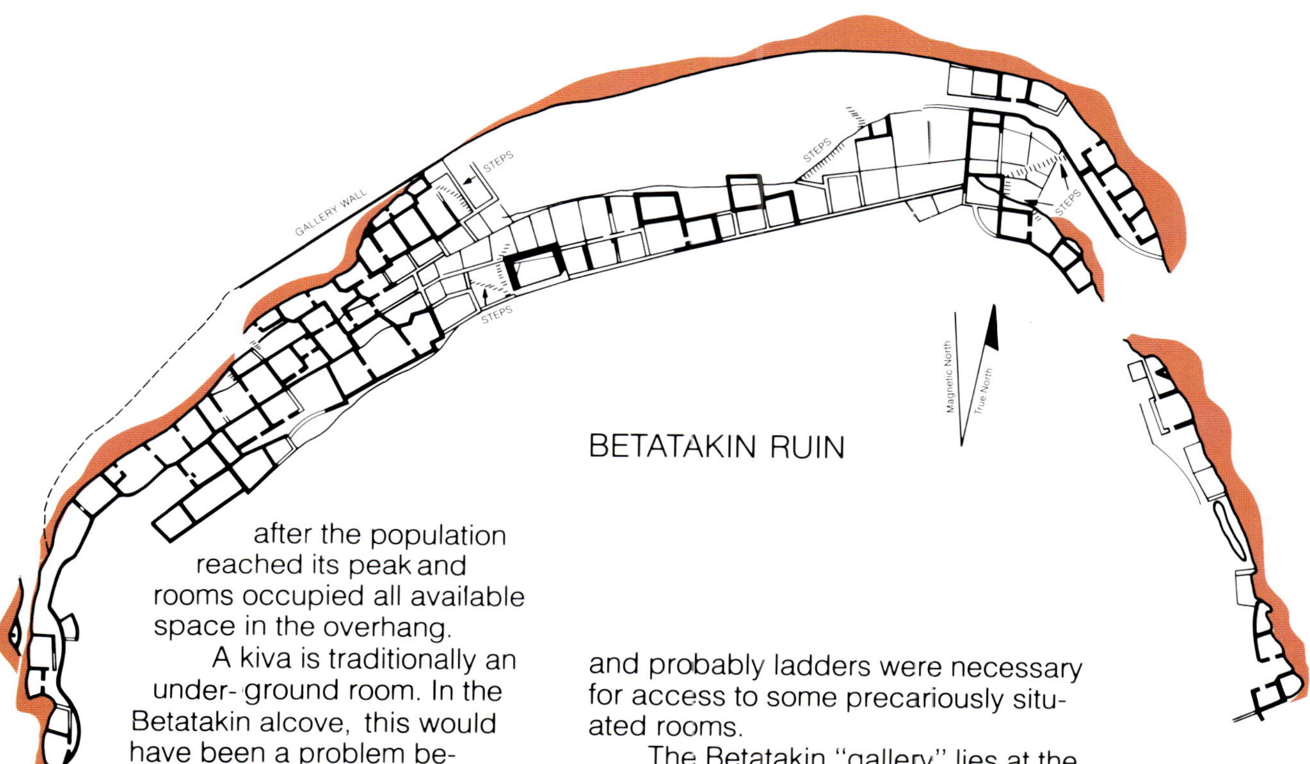

BETATAKIN RUIN

after the population reached its peak and rooms occupied all available space in the overhang.

A kiva is traditionally an under-ground room. In the Betatakin alcove, this would have been a problem because of the solid bedrock. The Anasazi got around this by making the kiva walls a double thickness of masonry, allowing them to simulate the effect of an underground chamber.

The steep pitch of the bedrock floor made building difficult in some areas of the alcove. The Anasazi used large amounts of mud mortar in wall construction to literally "glue" the rooms to the sloping bedrock. In some places, they pecked notches in the sandstone to provide a better foundation for walls. Toe and hand hold trails and probably ladders were necessary for access to some precariously situated rooms.

The Betatakin "gallery" lies at the rear of the alcove, some 9 meters (30 feet) above the village. This structure consists of a low masonry wall built along a high ledge. Food was probably stored here. It is an ideal spot, inaccessible to rats and mice and well protected from damaging moisture.

How did the Anasazi gain access to the gallery? They might have climbed up the long Douglas-fir pole that reaches up to the ledge from the floor below, though this would have been a difficult way to carry loads up. More likely, there was a ladder leading up from one of the now collapsed rooftops.

likely, there was a ladder leading up from one of the now collapsed rooftops.

The Betatakin Anasazi left rock art on the sandstone walls east of the main overhang. The pictographs include—in addition to the alleged Hopi Fire Clan symbol—a figure made up of concentric circles. Some Hopi visitors to the site claim that this symbol refers to the Hopi migrations, which were gradually coming closer to the center at the Hopi mesas. Just beyond are petroglyphs of several bighorn sheep. Could these have been part of a hunting ritual?

Hopi symbol—snake messenger.

The settlement, growth, and final abandonment of Betatakin all took place in less than two generations. The Anasazi left sometime after A.D. 1286, when the last beam was cut. There is no evidence of violence or hasty departure. The people probably moved out of the village in the same leisurely fashion that they moved in. By about A.D. 1300, the alcove in Betatakin Canyon once again stood silent and empty.

Keet Seel

The Anasazi who founded Keet Seel began their new village around A.D. 1250. They chose to locate about 8 kilometers (5 miles) upstream from where Keet Seel Creek joins the main Tsegi drainage. Here they found a large recess in the Navajo sandstone, situated at the head of a rincon on the west side of Keet Seel Canyon. Like the alcove at Betatakin Canyon, this spot offered shelter, ample water, and good farmland.

The people who moved into the canyon in the 13th century were not the first Anasazi to make use of the rock shelter. When they arrived at the site, they found the remains of at least two previous Anasazi groups who had also taken advantage of the location. These former occupations both date back to earlier Pueblo times. The new settlers razed whatever structures were not already ruined in order to clear space for their own buildings. Some of the old beams they used in their new houses, others they burned for firewood.

Keet Seel grew slowly at first. For the first 12 years, families arrived sporadically. Then between A.D. 1272 and 1276, there was a surge of tree-cutting activity and construction—indicating a major influx of people. During this period, the village took on its present appearance. At least five new families built homes in the village and the Keet Seel population reached about 80 people.

Keet Seel eventually expanded to a town of 125 to 150 people. Over 150 rooms stretched across the long upper floor of bedrock, and several more rested on the alluvium below. As the alcove filled with rooms, the Anasazi must have felt pressed for space to expand; two of the latest rooms perch on a hard-to-reach section of bedrock in the south-central part of the overhang.

For the last 20 years of the Keet

Because of its southeasterly exposure, the Keet Seel overhang offers a welcome retreat from heat of the summer sun. In winter the low sun provides natural solar heating.

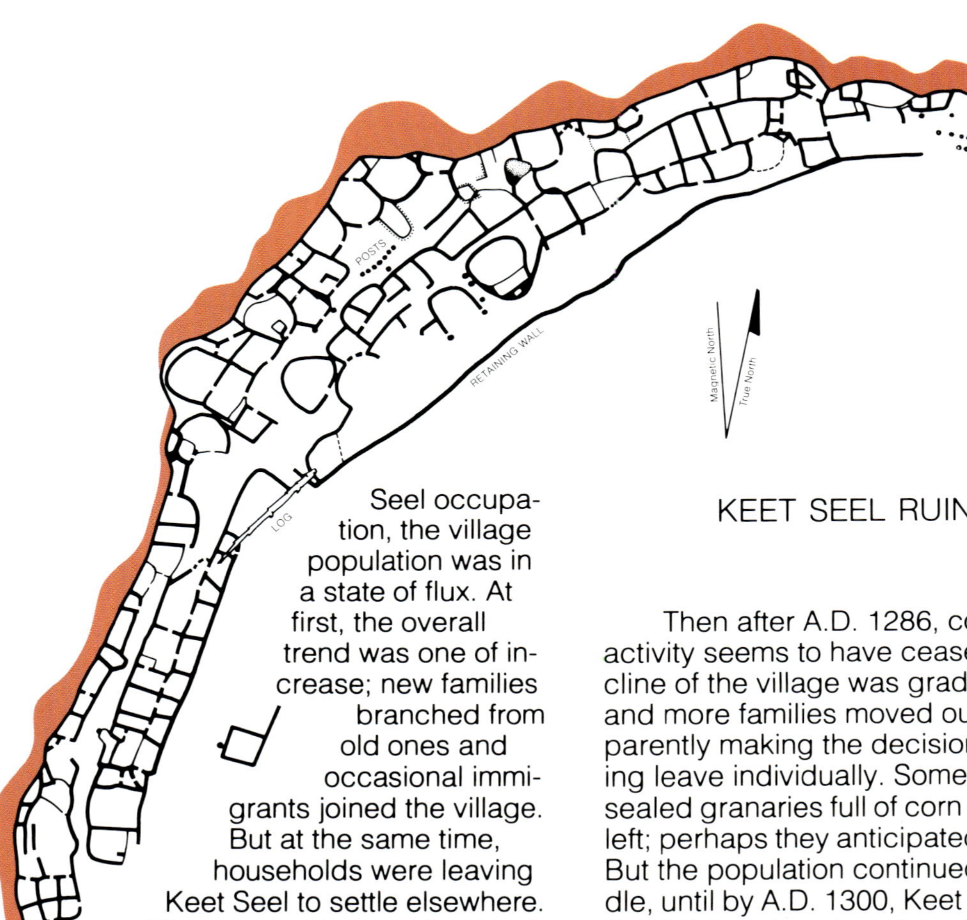

KEET SEEL RUIN

Seel occupation, the village population was in a state of flux. At first, the overall trend was one of increase; new families branched from old ones and occasional immigrants joined the village. But at the same time, households were leaving Keet Seel to settle elsewhere. The people who remained took beams from abandoned houses to put into new buildings. Newcomers occupied vacant homes and made them over for their own use.

The Anasazi converted many unoccupied dwellings into granaries. Was food already in short supply? If so, perhaps the people were storing reserves in anticipation of future shortages.

Then after A.D. 1286, construction activity seems to have ceased. The decline of the village was gradual. More and more families moved out, each apparently making the decision and taking leave individually. Some people sealed granaries full of corn before they left; perhaps they anticipated returning. But the population continued to dwindle, until by A.D. 1300, Keet Seel — like Betatakin — lay abandoned.

Whereas at Betatakin we see a planned move by a single group of people, the story of Keet Seel is different. The Anasazi who settled Keet Seel were a more diverse assemblage. Families arrived and departed independently. The tight village organization found at Betatakin does not seem to have existed at Keet Seel.

The people of Betatakin and Keet Seel all shared cultural patterns recognizable as Kayenta Anasazi. But because Keet Seel residents came to the Tsegi from a number of different communities, they showed more variations on the common Kayenta cultural themes. Turnover of Keet Seel households was common, so they did not have time to evolve more standardized patterns.

Building techniques at Keet Seel, though typical of the Kayenta area, show more diversity than those at Betatakin. Each Anasazi group arrived at the village with its own local building style. The quality of the work is generally poorer than at Betatakin; perhaps there was less social control to enforce building standards.

Though steep and difficult to reach, this section of bedrock may have been one of the last available building spaces in the Keet Seel rock shelter. The two rooms here were probably storage chambers.

The number and variety of kivas at Keet Seel is further indication of the heterogeneity of the population. The number of ceremonial groups was apparently greater here than at Betatakin. The Anasazi built four kivas at Keet Seel, no two quite the same shape. In addition, there are two isolated kivas up

In this keyhole-shaped kiva, the Keet Seel Anasazi held ceremonies to bring rains and make their crops grow.

the canyon in Turkey Cave that they probably used, too.

The keyhole kiva, in the central part of the lower eastern street at Keet Seel, is especially well preserved. Smooth sandstone slabs line its floor. A long, rectangular pit lies just north of the fireplace. Some archeologists suggest that, during ceremonies, this concealed a man who performed magical feats to heighten the effect of the ritual. Others think the Anasazi placed a wooden plank over the pit and danced on it as a foot drum.

The remains of several coats of plaster cover the interior kiva walls. The Anasazi painted two broad, white bands around the room on the paster. These match similar bands painted on the front of the two granaries above the

"Though the Anasazi who settled Keet Seel were a diverse assemblage, there must have been some over-all village organization. Several construction projects required the consent and active participation of most of the village households."

This D-shaped ceremonial annex lies adjacent to the keyhole kiva at Keet Seel. A wattle and daub wall such as this one is unusual in an Anasazi ceremonial room.

kiva. This suggests that the kiva and granaries are related. Did the people store special ceremonial corn in these two granaries?

Keet Seel has two rooms identified as "ceremonial annexes," each closely associated with one of the formal kivas. The annexes are D-shaped, with the straight wall of wattle and daub construction. These structures differ from domestic rooms, but do not have all the features of true kivas. They apparently provided additional space for ceremonial observances—perhaps minor ones.

In spite of the diversity of the Keet Seel population, there must have been some over-all village organization. For Keel Seel to function as a village at all, there had to be at least loose social ties and expectations to bind the residents together. Several construction projects required the consent and active participation of most village households.

One of these projects is the great retaining wall—up to 3 meters (11 feet) high in some places—which stretches 54 meters (180 feet) across the front of the eastern half of the village. Behind it lie several tons of fill. This impressive engineering feat is a credit to Anasazi ingenuity! To accomplish such an enormous task required involvement of the whole village. Some type of village organization had to exist to enable the people to decide on and plan the wall, and then mobilize the manpower to build it.

Construction of the Keet Seel streets required intra-village cooperation, too, though not on the scale of the retaining wall. Formal streets are unusual in an Anasazi cliff dwelling, but Keet Seel has three. One runs on the fill above the retaining wall; another sits on bedrock and fill in the rear of the eastern part of the alcove. The third street extends through the southwestern end of the village. Rooms line these long and relatively wide streets, which connect different parts of the village.

The Keet Seel Anasazi also built toe and hand hold trails, which are

One of the Keet Seel streets runs the length of the eastern end of the village above the retaining wall. Kivas on its northern side, when complete, would have extended out into the street and made it narrower than it now appears.

more typical in Anasazi sites. They pecked some of these into sandstone in the central part of the alcove to provide access to the main village above. Today visitors climb up a ladder built in the same area by the National Park Service to accommodate modern visitors — often not as agile as the Ancient Ones! Other toe and hand hold trails in the sandstone cliffs near the village provided routes onto the mesa above, where the Anasazi gathered firewood and hunted game.

A large, suspended log — embedded in masonry at either end — sits above the access ladder. The log is a Douglas-fir tree, which the Anasazi chopped down with a stone axe and hauled from the far side of the canyon where the closest Douglas-fir grow. What did they put it there for? To go to so much trouble, they must have had something special in mind. Could they have used it to hoist materials up to the village on ropes?

Most of the rock art at Keet Seel is concentrated on the roof of the overhang just west of the big log. Here vil-

The purpose of this large log at Keet Seel is a mystery.

lagers depicted game animals, human figures, and what appear to be strange, turkey-like birds on top of human bodies. Interpretation of the pictographs is open to speculation.

"It seems likely that the canyon channeled travel and communication between the Tsegi villages, so that they formed a loose community."

Nearly half a century of human occupation left an accumulation of garbage and used and broken belongings in the village. We can see places where the Anasazi built walls on top of refuse piles. Much of the fill banked behind the retaining wall and buried beneath the street is trash. Villagers also tossed garbage over the edge of the upper terrace, creating a large trash mound stretching along the base. Visitors today can still see this refuse: pieces of broken pottery, fragments of yucca fiber rope, turkey feathers, stone scraps from tool-making, corncobs, squash rinds, and many other discarded items of an ancient culture.

The remains of several rooms lie on the trash and alluvium below the main village, though erosion has destroyed most of the evidence in this lower occupation area. One of these structures is especially intriguing. It appears to be a round, two-story tower. This type of building is common in the Mesa Verde area, but extremely rare at Kayenta sites. Perhaps someone from Keet Seel brought the idea back from a trading expedition to Mesa Verde. Or some Anasazi people from Mesa Verde may have settled at Keet Seel and built the tower themselves. Who knows what other goods and ideas were included in such an interchange?

Were the Keet Seel Anasazi in contact with their neighbors at Betatakin? There is no evidence of formal intervillage coordination between Keel Seel and Betatakin or any of the other, smaller settlements in the Tsegi. The sites are close enough, though, that we can assume there was interaction between them. No doubt they exchanged trade goods and information. There was probably a good deal of intermarriage between the villages, too. It seems likely the canyons channeled travel and communication between the Tsegi villages, so that they formed a loose community.

Inscription House

Inscription House sits on a ledge part way up the southern face of a huge sandstone dome on the northern side of Nitsin Canyon. A shallow overhang in the rock face protects the rooms from all but the most severe rainfall. The site lies only .8 kilometers (a half mile) from where Nitsin Canyon joins the main Navajo Canyon drainage. Ample farmland lies below the village in Nitsin Canyon and downstream at the junction of Nitsin and Navajo Canyons.

Unfortunately, we cannot trace the founding, growth, and abandonment of Inscription House with the same detail and accuracy that we can that of Betatakin and Keet Seel. Presumably independent households came together from numerous different locations to settle the village. Family groups started to arrive and build homes around A.D. 1250.

Inscription House sits high up in a great sandstone dome in Nitsin Canyon.

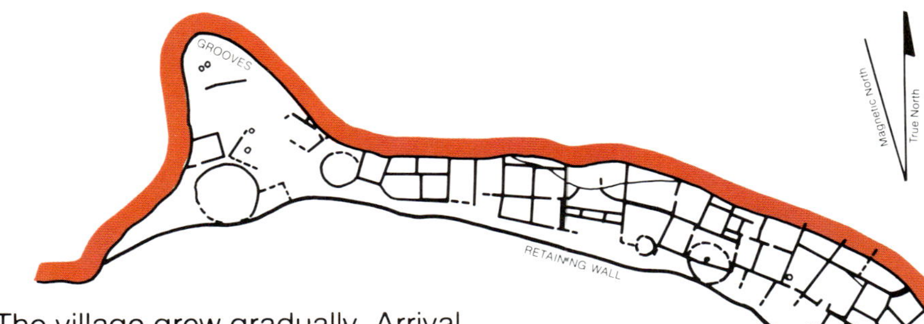

STEPS

INSCRIPTION HOUSE RUIN

The village grew gradually. Arrival of occasional later immigrants plus natural increase by original residents as a result of birth and marriage brought the population to a peak of about 75 people. Like the Anasazi at Betatakin and Keet Seel, the people of Inscription House left their village to move elsewhere by A.D. 1300.

A retaining wall runs along the front of most of the village. Almost all of the 80 rooms stretch along the level area provided by the fill banked behind the wall. The Anasazi did not occupy all the rooms at once. There is considerable evidence of rooms being torn down and remodeled. As households moved out and population waned, the remaining people dumped trash into many abandoned rooms.

"An interesting and unique feature of Inscription House is the use of adobe bricks. This type of wall construction is unknown in the Tsegi, but common in the Navajo Canyon region."

The occupants of Inscription House built only one recognizable kiva. Apparently the village participated as a whole in ceremonies. The kiva sits detached from other rooms, at the far western end of the village. To achieve the desired underground effect, the Anasazi cut and leveled the sloping bedrock at the edge of the overhang ledge. They pecked the rock to even out the surface, and then plastered floor and wall to obtain a smooth surface. To construct the upper walls of the kiva, they used the thin sandstone slabs removed in forming the floor.

A deep recess in the western end of the alcove, behind the kiva, contains no rooms. A few grooves in the rock at the very rear of this recess show that stone axes were sharpened here.

An interesting and unique feature of Inscription House is the use of adobe bricks in wall construction. Though much of the village has typical Kayenta stone masonry and wattle and daub, the Anasazi built almost half of the rooms totally or partially with adobe bricks. This type of construction is unknown in the Tsegi, but is common in the Navajo Canyon region. Did the Inscription House Anasazi perhaps use adobe bricks because sandstone blocks for building were not as easily found here as in the Tsegi?

They shaped the individual bricks by hand. Mud was applied to each side of a reinforcing core of bunched grass. The finished block dried in the sun. A few bricks have especially flat, smooth sides; these may have been formed in some kind of mold. After setting the bricks in place, the builders plastered the wall. This smoothed the surface and helped bind the bricks together; it also thickened the wall to provide greater strength.

Some of the adobe walls have T-shaped doorways. This type of door, like the adobe construction itself, appears in the later rooms. The Anasazi also built T-shaped doorways in the Tsegi — in particular at Keet Seel; but such entranceways occur more frequently at Inscription House. One second-story room with a T-shaped door has a matching T-shaped smoke-hole just above and to the right of the entrance!

The purpose of this shape is a mystery. The Anasazi may have found it easier to enter the room when carrying loads on their backs. Or perhaps they

Most rooms at Inscription House sit on the fill banked behind the front retaining wall to level the steep bedrock floor.

hung a piece of hide over the upper, wider part of the door for warmth, and still had a small opening for ventilation at the bottom. Or people may have used the projections for balance when stooping to enter small doorways. Other explanations are possible.

An extensive trash mound lies on the slope at the cliff base below Inscription House. Refuse accumulated here as the Anasazi threw it down from the village above. In addition to the cast-offs of Inscription House dwellers, the midden has revealed potsherds dating as far back as about A.D. 600. Apparently early Anasazi groups occasionally used the recess for temporary shelter. In addition, they occupied the alcove on a full-time basis during the last half of the 12th century. When the founders of Inscription House arrived, they must have cleaned out the trash left behind by earlier occupants and tossed it onto the slope below.

"A comparison of Inscription House, Keet Seel, and Betatakin can teach us much about local variations within a single cultural tradition."

The Inscription House Anasazi buried their dead in the trash midden. Archeologists found 32 burials here — considerably more than the few uncovered at Betatakin and Keet Seel. Most of these included some kind of mortuary offerings. The most elaborate was the grave of a woman in her late 40's, which included 12 pottery vessels, a metate and mano, a pair of turquoise pendants, a stone maul, and numerous bone tools.

The variety in the number and kinds of burial offerings — especially among adult women — may reflect differences in status. Perhaps the woman mentioned above was a member of a prestigious family. Status could also be recognition of excellence in some specialized skill, such as pottery-making.

Though located 48 kilometers (30 miles) northwest of the towns in Tsegi Canyon, Inscription House was contemporary with Betatakin and Keet Seel. Archeologists recognize all three villages as part of the same phase of Kayenta Anasazi development. Their residents shared a similar culture and material technology, but at the same time each community did some things a little bit differently from people at the other villages. A comparison of Inscription House, Keet Seel, and Betatakin can teach us much about local variation within a single cultural tradition.

Architectural differences — such as use of abode bricks and frequency of T-shaped doorways at Inscription House — are just one example. Perhaps there were also slight variations in pottery styles or in other things that the archeological record tells us nothing about — such as stories and ceremonies. Even within the general framework of Kayenta Anasazi culture, we should expect to find local differences from area to area and village to village.

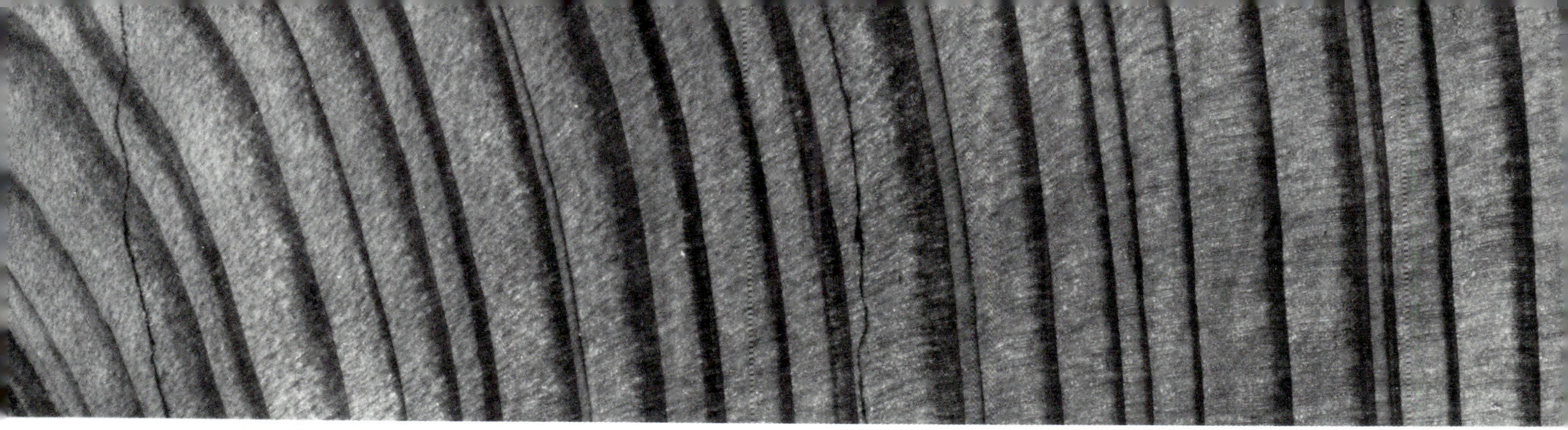

RECONSTRUCTING PREHISTORY

"It seems remarkable that archeologists are able to speculate on the complexities of human behavior using such simple evidence as buildings, tools, and potsherds."

How have archeologists come to understand so much about the prehistoric Anasazi based on what seems like so few clues? It seems remarkable that they are able to speculate on the complexities of human behavior and society using such simple evidence as buildings, tools, and potsherds. The progress and development of archeological techniques is in itself a fascinating subject. Southwestern archeology has come a long way since its beginnings in the last half of the 19th century.

The first archeologists who dug in the Southwest did not know when the ruins that they found had been occupied. They could get a relative idea of the age of features and artifacts, though, by studying the stratigraphy or layering of a site. For example, we can assume that things found on the bottom of an undisturbed trash mound are the oldest, and those lying on top the most recent. This is because people usually toss their current garbage on top of what they have thrown out before.

The development of dendrochronology — or tree-ring dating — gave researchers a more accurate and absolute way to date sites. By taking a core sample from a beam found in a ruin and matching it with a master tree-ring chronology for the area, a dendrochronologist can determine in what year that tree was cut. He can sometimes even tell whether it was cut in the spring or fall of the year. This is the most exact method of dating available to archeologists today.

In addition, the width of growth rings indicates wet and dry years. Analysis of pollen grains collected at a site also provides clues about the prehistoric climate. By finding out which plant species were common in the area at the time, researchers form an idea about the temperature and rainfall which supported that type of vegetation. These factors also affect growing conditions for crops, an important consideration in the study of an agricultural people.

Southwestern archeologists show an interest in pottery that goes beyond pure aesthetics. The study of prehistoric ceramics is an important subdiscipline of archeology. Potsherds — pieces of broken pottery — are virtually indestructible and occur in abundance at most Anasazi sites.

Dendrochronologists take core samples from a beam in order to determine when the tree was cut. Tree-ring dating enables archeologists to accurately trace the growth of prehistoric towns like Betatakin and Keet Seel.

By studying these sherds they learn about styles and techniques and how these changed through time. Based on such characteristics as materials and design, they have identified and labeled hundreds of different pottery types with names like Moenkopi Corrugated, Kayenta Black-on-White, and Keet Seel Polychrome. Carefully noting which pottery types are present tells a researcher much about who occupied a site, during what period, and about what other groups they were trading with.

Early archeologists tended to concentrate on the study of Anasazi material culture, such as pottery, baskets, and arrowheads. Many were after especially fine specimens for display in museums. Their reports usually consisted of lists and descriptions of the material traits of a site. They were trying primarily to date and classify different Anasazi groups, thus helping establish foundations and framework for later research.

Archeologists only recently have turned serious attention to investigation of prehistoric social organization. Now they discover that by observing location and distribution of artifacts at a site, they can learn much about the society of the people who used them. Ethnographic inference also helps; researchers gain insight by comparing prehistoric evidence with the living societies of groups like the Hopi.

More and more, archeologists are becoming interested in not just the

The use of adobe brick and the frequency of T-shaped doorways distinguish Inscription House Anasazi from their contemporaries in Tsegi Canyon.

"More and more, archeologists are becoming interested in not just the 'what?' and the 'when?' of Anasazi culture, but also the 'why?'."

"what?" and the "when?" of Anasazi culture, but also the "why?." Today many are especially interested in the relationship between prehistoric man and his environment. They want to know how the land and climate influenced development of such cultures.

"We must remember, though, that archeology — like any other science — is based on theory. Ideas are constantly changing."

The growth of contract or salvage archeology is helping expand the focus of archeological research. Contract archeologists gather evidence from sites threatened by construction of dams, roads, and power plants. Salvage work has directed attention to the less spectacular but still important sites that most early archeologists passed over. The information gained has helped place the larger ruins like Betatakin and Keet Seel into a broader picture.

Modern archeologists have at their disposal sophisticated methods undreamed of by their predecessors. New techniques such as archeomagnetic dating, remote sensing, and the use of computers and statistical analysis are helping to further our understanding of the workings of prehistoric cultures. Specialists from other fields such as geology, botany, and zoology are contributing their insights, too.

We must remember, though, that archeology — like any other science — is based on theory. Ideas are constantly changing. New data may support a current theory, or it may disprove that theory and plant the germ for yet another. We will never know the whole story of the Anasazi and the life they led, but the learning process is exciting and challenging. Betatakin, Keet Seel, and Inscription House still have much to reveal about their builders.

DISCOVERY & EXPLORATION

"The remote and rugged region surrounding Betatakin, Keet Seel, and Inscription House was one of the last in the country penetrated by the white man."

The remote and rugged region surrounding Betatakin, Keet Seel, and Inscription House was one of the last in the country to be penetrated by the white man. The first evidence of the presence of non-Indians in the area is the inscription for which Inscription House ruin is named. Great controversy has surrounded the identity of the date and the now illegible words scratched on the interior wall of one of the ruin rooms.

For a long time, experts believed the date to be "1661." They thought a 17th century Spanish missionary or explorer had happened upon the deserted village and inscribed the figures into the wall plaster. More recent investigators, however, have suggested the date originally read "1861." The discovery of two other names elsewhere in the canyon along with the date "1861" seems to substantiate this theory.

A search of Mormon historical records revealed that these names belonged to two men from Cedar City, Utah. In 1861 they joined the Mormon missionary Jacob Hamblin on an expedition across the Colorado River to recover the body of George Smith, Jr. Several months before, hostile Navajos had killed young Smith, who was accompanying Hamblin on a mission to the Hopi mesas.

Richard, the oldest Wetherill brother, devoted much of his life to searching for Anasazi ruins. He discovered Keet Seel in 1895.

The Mormon expedition could have passed through Navajo Canyon on the way to the Colorado. Members of the party probably climbed up into the ruin in Nitsin Canyon and left the name and date. It seems likely that these Mormons, rather than a 17th century Spaniard, were the first white men to see Inscription House.

Credit for reporting and publicizing the ruins which would become Navajo National Monument goes to the Wetherill family. The five Wetherill brothers were originally ranchers from Mancos, Colorado. Though they had no formal training in archeology, they had an avid interest in the Anasazi and the remains of the ancient culture. The Wetherill brothers are responsible for the discovery of many major Anasazi sites in the Four Corners area — including most of the large cliff dwellings at Mesa Verde.

In January, 1895, the oldest brother, Richard, accompanied by Charlie Mason, rode through Marsh Pass and started up into the Tsegi. When they reached the spot where the main canyon branches, they chose to continue to the right up the center fork. Thus they came eventually upon Keet Seel, and became the first white men to see the crumbling remains of the Anasazi village. Thousands of potsherds

scattered throughout the site justified the Navajo name given to the ruin — Keet Seel or "broken pottery."

Wetherill and Mason spent several days exploring the large ruin, but undertook little excavation. Richard returned two years later as guide for an expedition financed by a wealthy Harvard student, George Bowles. This party worked more extensively in the ruin.

The group had completed its excavation at Keet Seel and was camped at Marsh Pass, when Indians kidnapped Bowles and his tutor and held them for ransom. Richard dispatched a rider to Bluff, Utah to bring back the money, and upon receiving the silver, the Indians released their captives. Bowles and his tutor returned to camp, shaken but unharmed.

John, the second Wetherill brother, worked as an Indian trader at the present day site of Kayenta, about 32 kilometers (20 miles) east of Tsegi Canyon. For 30 years his trading post served as the base of operations for archeological exploration in the Marsh Pass–Tsegi Canyon area. John himself was guide and outfitter for many of the expeditions. He later served as the first custodian of Navajo National Monument.

John Wetherill's warm hospitality and expertise as a guide drew many travelers and professional researchers to his trading post in Kayenta.

In July of 1909, John Wetherill led Byron Cummings — an archeologist from the University of Utah — and a group of his students into Nitsin Canyon, where they found Inscription House ruin. Two children — the son of Cummings and the daughter of Wetherill — discovered the mysterious, faded inscription, partially concealed by debris from a fallen roof. Cummings was to return many times in later years to excavate in the ruin.

Later that same summer Wetherill and the Cummings party — following directions given to Mrs. Wetherill by an old Navajo woman — came upon the ruin of Betatakin for the first time. They had found the third and final cliff dwelling that would make up Navajo National Monument. The Navajo name for the site — Betatakin or "ledge house" — refers to the steepness of the bedrock floor on which the village sits.

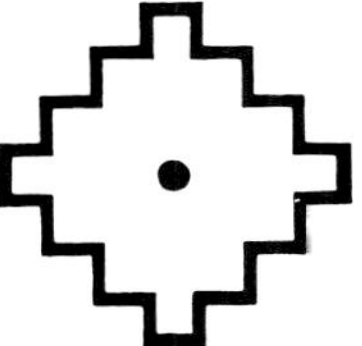

Hopi symbol — home of the Spider Woman symbolizing unfailing wisdom.

Cummings went back to Betatakin in September with his University of Utah students and spent three months excavating the site. Neil M. Judd, who had been among Cummings' students, returned to the ruin in 1917 under the auspices of the Smithsonian Institute in Washington, D.C. He worked several months on stabilization and repair of the prehistoric village. Judd and his crew did an admirable job, in spite of such hardships as the severe weather and the difficulty of obtaining even basic supplies in such a remote location.

Judd had originally planned to work at Keet Seel, too, but lack of sufficient time and money forced him to concentrate his efforts on Betatakin. A Civil Works Administration crew, under direction of John Wetherill and Irwin Hayden, finally stabilized Keet Seel in the 1930's. Much of their work was reconstruction of the crumbling retaining wall, which was badly in need of repairs. They also undertook some excavation in rooms which had not been dug and in the trash below the village.

Many men whose names are famous in Southwestern archeology passed through and viewed the three cliff dwellings. Only a few, like Cummings and Judd, stayed long enough to do any real work. Unfortunately, in spite of the fame and archeological importance of the sites, little information on them exists in print. Cummings never wrote the results of his work at Betatakin and Inscription House, and a garage fire at his Tucson home in the 1940's destroyed most of his field notes.

In the early 1960's Jeffery S. Dean, of the University of Arizona Laboratory of Tree-Ring Research, spent several field seasons at the cliff dwellings in Tsegi Canyon. His thorough study of tree-ring dates at the sites has given us a great deal of invaluable information about the development of Betatakin and Keet Seel and about the Anasazi in the Tsegi.

Recent work at Inscription House has helped archeologists fill in some details of the Anasazi occupation there. The Museum of Northern Arizona conducted a National Park Service sponsored excavation at the site in 1966. Further digging in conjunction with National Park Service stabilization of the ruin in the summer of 1977 will add more information in the near future.

PROTECTION & PRESERVATION

"Today the isolation factor can no longer be counted on to protect the cliff dwellings. Human erosion has become a problem."

A presidential proclamation in March, 1909, established Navajo National Monument. This was several months before anyone knew of the existence of Betatakin and Inscription House! The original boundaries included a much larger area than the present monument now encompasses. A second proclamation in 1912—based on further surveys and recommendations by experts who visited the area—reduced the monument acreage. It now includes just the three major ruins—Betatakin, Keet Seel, and Inscription House—and the land immediately surrounding them.

At first visitors to the monument were very few. Few people except professionals undertaking scientific study knew of the ruins and were willing to make the arduous trip to visit them. Not until the 1930's had travel to Navajo National Monument increased enough to require a permanent headquarters and a full-time resident custodian.

This extreme isolation did not last. In 1959 construction began on a cross-reservation highway, and in 1965 the monument entrance road was completed. Paved roads now connected Navajo National Monument with the outside world. Visitation shot up almost immediately from 1000 people per year to about 1000 people per month. The National Park Service built a new visitor center, campground, and trail to Betatakin to accommodate the new influx of visitors to see the ruins. Today some 50,000 people come to the monument each year.

For many years difficult access to the cliff dwellings kept visitor traffic to a minimum. Damage to the fragile ruins from visitor use was not much of a problem; nor was the illegal removal of potsherds and other artifacts remaining at the sites. But today the isolation factor can no longer be counted on to protect the cliff dwellings. Human erosion has become a problem.

In an attempt to safeguard the sites against carelessness and overuse, the National Park Service instituted guided tours through the cliff dwellings and closed especially fragile sections of the sites to the public. Finally, a limit was set on how many people could visit the ruins. By these measures, the National Park Service hopes to preserve these valuable archeological sites for the use and enjoyment of future generations.

Employees of the National Park Service periodically do minor stabilization and repair work at Betatakin, Keet Seel and Inscription House. Because of the exceptional state of preservation of the cliff dwellings at Navajo, they feel a commitment to keep the sites in as natural a state as possible. In their repair work, they have tried to use the original materials and construction techniques. They have tried to stabilize rather than reconstruct—only to rebuild as much as is necessary to maintain the ruins.

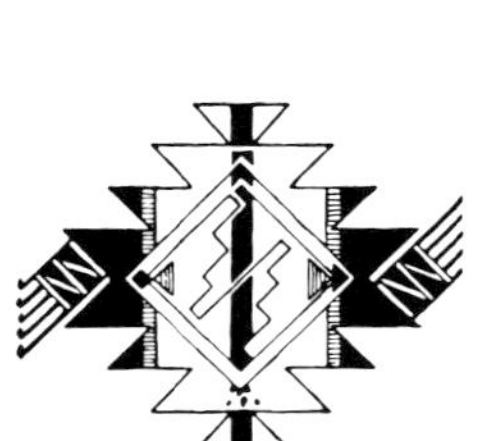

Hopi symbol—prayer for good life (from design on wedding blanket).

The National Park Service considers it more important to preserve the cliff dwellings in an original condition than to enable them to withstand heavy visitation through artificial means, such as cement and steel reinforcements.

The trail to Betatakin includes some 720 steep, limestone steps—about the equivalent of climbing a 70-story building.

Today the visitor to Navajo National Monument does not have to undergo the risks and discomforts of travel on rough, often impassable dirt roads. But the monument still lies in a remote and undeveloped area, miles from many of the amenities of modern civilization that most of us take for granted. To see and fully appreciate the ancient Anasazi villages, the visitor has to spend a little time and expend some effort. He must leave his vehicle behind and set out into the canyon country on foot.

An easy 1.6 kilometer (1 mile) round trip walk along the rim of Betatakin Canyon will reward the visitor with a distant view of Betatakin ruin, set deep within the huge sandstone arch. If he is willing to spend a morning or an afternoon and walk 214 meters (700 vertical feet) down the steep steps into Betatakin Canyon, he can experience first hand the verdant lushness of the narrow canyon and stand within the great arching vault where the Anasazi built their village 700 years ago. A 48 kilometer drive—the last part still unpaved—plus a 4 kilometer (2.5 mile) hike, takes the visitor through slickrock and canyon country to Inscription House ruin, perched high on a ledge overlooking Nitsin Canyon.

Visitors hike into Nitsin Canyon on their way to Inscription House.

Probably the ultimate experience at Navajo National Monument is reserved for the most ambitious visitor. To reach Keet Seel ruin, he must undertake a rugged 10 kilometer (6.5 mile) hike or a 13 kilometer (8 mile) horseback trip through the steep sandstone canyons of the Tsegi. For his effort, he can see and explore one of the largest—and certainly the best preserved—

"The visitor sees the canyon country and it takes on a new dimension when he views it not just as pretty scenery, but as natural forces that shaped the daily lives of Anasazi people."

cliff villages in the Southwest. Its remoteness, size, and remarkable preservation give Keet Seel a special quality. It seems hard to believe that it has already been 700 years since Anasazi people made their home here.

The ruins, the story of the Anasazi, and the surrounding land are all part of Navajo National Monument. The visitor hears the archeologists' theories about the cliff dwellers, and uses his imagination to fill in the gaps, adding the personal, human touches that theory cannot provide. He sees the canyon country, and it takes on a new dimension when he views it not just as pretty scenery, but as natural forces which shaped the daily lives of the Anasazi people. Hopefully the visitor will come away with some sense of the life led by a people 700 years ago — a life in many ways simpler than his own. Perhaps he can imagine what it was like to live in more intimate association with the land than most of us do today.

NAVAJO: TRADITION & CHANGE

> "Though the Navajo did not build or occupy the cliff dwellings, it is their culture that gives the area its distinct flavor today."

The Navajo Indians and their way of life form an important part of any visit to Navajo National Monument. Though the Navajo did not build or occupy the cliff dwellings, it is their culture which gives the area its distinct flavor today. The Navajo refer to themselves as Dineh — "The People." With some 150,000 members, their tribe is the largest and fastest growing in the United States today. Their reservation — also the largest in the country — covers 64 million hectares (16 million acres) and includes parts of three states — Arizona, New Mexico, and Utah.

In comparison with most American Indian tribes, the Navajo have preserved much of their tradional culture. Their history shows them to be an adaptable people. Over the years, they have learned many things from other groups with which they have come in contact — the Pueblo Indians, the Spanish, and the Anglos. They have succeeded in integrating these acquired ideas into their own culture, and giving to them a distinctly Navajo character. This ability to blend the new with the old has enabled them to change and at the same time retain their own unique culture.

The Navajo are Athabaskan speakers. Their closest linguistic relatives — other than the Apache — live in northern Canada near the Arctic Circle. Anthropologists believe they came originally from this northern region. They entered the Southwest between 1300 and 1500, and first settled in the area of northwestern New Mexico known as Dinetah — or "Old Navajoland."

The Navajo people brought with them the hunting and gathering ways of the far north. In the Southwest, they encountered the Pueblo Indians, who had occupied the area for centuries. From these peaceful farmers, they learned to till the land and grow their food.

Arrival of the Spaniards in the 16th century had a profound effect on the lifeways of the Navajo people. Spaniards brought the first domesticated sheep, goats, cattle, and horses to the Southwest. The Navajo acquired sheep and goats, and herding became a central part of their livelihood. The sheep provided meat and wool; the size of a person's herd was a measure of his wealth.

The Navajo did not care to settle in villages like those of the Pueblo people. They adopted a pastoral way of life more in keeping with their traditional wandering ways. Their existence revolved primarily around the needs of their animals. Families moved seasonally in search of good grazing and water for their sheep.

In winter they lived in relatively permanent settlements. They constructed winter houses — called "hogans" — of logs, stone, and mud. The traditional hogan is round, with the door opening to the east toward the rising sun. In

A Navajo family travels by wagon in the early part of the century.

summer the people shifted locations frequently, following their herds. They stayed in temporary brush shelters that were easy to build and abandon.

A cluster of hogans made up a Navajo settlement or "camp." An extended family lived in each camp — usually an older couple, their daughters, and their daughters' families, each in their own house. The Navajo — like the Anasazi and the Hopi — traditionally trace descent through their women, and, when they marry, live with the wife's family.

Though each sheep belonged to an individual, the extended family cared for the flock as a whole. Young children were usually responsible for herding.

A married couple often planted a field in the vicinity of the camp. They grew corn, beans, watermelon, and squash, and sometimes raised peach trees, too. They supplemented their crops by collecting wild foods — such as pinyon nuts and yucca and cactus fruits — in season. Meat obtained by hunting game like deer and rabbit provided a change from roast mutton.

Raiding also formed an important part of the Navajo economy during this period. Introduction of the horse by Spaniards gave them a new mobility, and with it an easy way to add to the size of their herds. Bands of mounted Navajo warriors swept down on Pueblo towns and white communities — first Spanish and later American — seizing captives and livestock. They also raided to retaliate for injuries and to defend their territory against white encroachment.

In reprisal, the Spanish and then the United States military launched numerous expeditions against the Navajo. The U.S. government made treaties with some local Navajo "chiefs," but other Navajos continued their raids. The white men did not understand that a Navajo headman had authority only over his own local outfit, and that other groups did not recognize these treaties. Government officials viewed the Navajo as incurably treacherous.

Navajo girls herding sheep and goats in the early 1900's.

In 1863, the U.S. Army ordered the famous frontiersman Kit Carson to subdue them. Carson marched across the Navajos' homeland, attacking the people, burning their fields, and destroying their livestock. His campaign was devastating. Thousands of starving and defeated Navajos surrendered at Fort Defiance, to make the "Long Walk"— 482 kilometers (300 miles) to Fort Sumner, southeast of Santa Fe on the Pecos River in New Mexico.

Some Navajos managed to escape Kit Carson and the trip to Fort Sumner by hiding out in inaccessible areas, particularly in the western Navajo country. The Tsegi and other remote and little-known canyons around what is today Navajo National Monument sheltered many Navajo refugees. In fact, it was probably the flight from Kit Carson that first brought substantial numbers of Navajo people to live in this western region.

The U.S. government held some 8500 Navajos captive at Fort Sumner for four years. These proud and independent people had to depend on the government for their food and supplies. Many hundreds of them died of disease or starvation. Government agents tried to turn the Navajo into sedentary farmers; but the People — accustomed to ranging freely with their flocks over large areas of land — resisted. The confinement at Fort Sumner was a period of suffering and demoralization for the Navajo people.

Finally, in 1868, the federal government agreed to let the People return to their homeland. They allotted the Navajo some 14 million hectares (3.5 million acres), centered around Canyon de Chelly, for a reservation. The destitute Navajo received seeds, tools, some livestock, and — for a time — rations from the government. They had to rebuild their way of life from scratch. Gradually they re-established themselves, increased their flocks, and once again attained a degree of prosperity.

During their stay at Fort Sumner, they learned many things from their captors. They developed a taste for Anglo foods like coffee and flour. They adopted new styles of dress. The women modeled their long, wide skirts after those of the Army officers' wives. They topped these with brightly colored velveteen blouses. The men wore similar shirts with long trousers made from gifts of unbleached muslin that they received at the fort. The Navajo acquired tools and other manufactured articles at Fort Sumner.

Reservation trading posts were also important in introducing them to the white man's ways and goods. Here

"The ability to blend the old with the new has enabled the Navajo to change and yet at the same time to retain their own unique culture."

Until recently, horses were the primary means of transportation on the vast Navajo reservation.

Navajo weaving techniques are slow and laborious; even today, a weaver receives relatively little for the number of hours she puts into a rug.

they could see and obtain tools and utensils, bolts of bright cloth, clothing, and canned and packaged foods. For these ready-made items they traded wool, hides, livestock, and their crafts — primarily rugs and silver jewelry.

Traders advanced credit to reliable customers. Navajos frequently pawned jewelry, saddles, or rifles as collateral in exchange for the goods they needed. If they did not pay the debt and redeem their valuables, the trader sold the unclaimed pawn.

For the Navajo, the periodic trip to the trading post was more than just a business venture. It was one of the few opportunities to meet with friends and relatives and to exchange news and gossip. They could also hear of happenings in the white man's world beyond the reservation. The trader was the Navajos' primary contact with this outside world. Often people sought the trader to mediate disputes, give advice, or bury the dead.

Traders were instrumental in development of the Navajo rug. The People learned weaving techniques in the early days from the Pueblo Indians. In the case of the Navajo, however, the women do the weaving, rather than the men as among their Pueblo teachers. For a long time, Navajo women wove woolen blankets and dresses to wear and to trade with the Pueblos and Mexicans.

After Fort Sumner, traders helped create a broader market for Navajo weaving. By the end of the 19th century, Navajo women were producing coarse rugs, which were in demand across the country. For their own apparel they preferred the brightly colored Pendleton blankets made in Oregon and obtained from the traders. To encourage Navajo women to improve the quality of their work, many traders suggested attractive colors and new patterns. The sale of rugs still provides the Navajo with a source of income. Even with today's seemingly exorbitant prices, however, a weaver receives relatively little for the number of hours that she puts into a rug.

The Navajo learned to work silver from the Mexicans sometime between 1853 and 1858. In the years following Fort Sumner, they developed their silver and turquoise jewelry into a fine art. Their designs reflect influences from Spanish, English, and Arabic sources.

The Navajo prize their jewelry, not just as a valuable item for trade, but also as personal ornaments. Ceremonies and social gatherings provide the people with opportunity to display a wealth of silver and turquoise. Styles made by craftsmen of other Southwest Indian groups — such as the Zuñi — are also popular with them.

Sand paintings are a recent introduction to the Navajo craft market. Craftsmen base their designs on sandpaintings made in Navajo curing ceremonies. At these rituals the medicine man creates an elaborate pattern of symbols on the ground, using colored sands. He seats the patient on the finished work and performs certain prescribed actions, after which he destroys the painting.

The Navajo learned many religious concepts and practices from their Pueblo neighbors, though they adapted these ideas to fit their own cultural orientation. Like the Pueblos, they place emphasis on ritualism and a daily awareness of the supernatural. The Navajo, too, perform elaborate ceremonies with masked dancers.

"The Navajo did not care to settle in villages like those of the Pueblo people. They adopted a pastoral way of life more in keeping with their traditional wandering ways."

However, whereas Pueblo religion is concerned primarily with rain and fertility for farming, Navajo ceremonies focus mainly on curing illnesses. Pueblo rituals aim at maintaining the balance of the universe as a whole, while those of the Navajo attempt to restore harmony between man and the universe. The medicine man — or "singer" — cures his patient by performing a chant to ritually correct the imbalance believed to cause the sickness.

Navajos have a great fear of the dead. No matter how good a person was during his life, his ghost is a threat to the living. Ghosts appear after dark, and may chase people or make them ill. Because of this belief, Navajos will go out of their way to avoid contact with the dead, and will abandon a hogan if someone dies in it. No wonder people were so anxious to have the local trader bury their dead! The Navajo did not disturb the deserted Anasazi cliff dwellings for the same reason; they feared the ghosts of the ancient people who had lived and died in the stone villages.

During the late 19th and early 20th centuries, the U.S. government's policy towards the Navajo was to try to rid them of such "primitive superstitions" and to "civilize" them. Indian boarding schools removed children from their home environment — often under force — in order to teach them the ways of the white man, so that they could enter into American society. Unfortunately, most of these "educated" children found themselves unable to function in Anglo society, and were handicapped in returning to their own culture because they did not know the traditional skills and customs.

Recognizing these problems, the government altered its educational policy in the early 1930's; recent emphasis has been on teaching children the skills to enter into both Anglo and Navajo cultures. Most younger Navajos today speak both languages, though many of their elders are still solely Navajo speakers. There are two separate school systems on the reservation — boarding and day schools for grade school children administered by the Bureau of Indian Affairs, and state-run public schools at both the grade school and high school levels.

When the Navajo people returned from Fort Sumner in 1868, the U.S. government issued families small numbers of sheep and goats. They were very successful in building up their herds — perhaps too successful. Overgrazing coupled with natural drought reduced the plant cover and caused severe erosion throughout the Navajo country. By 1933, the problem had become so critical that the federal government instituted a stock reduction program. They tried to convince the People that if they decreased their flocks, the remaining animals would be of better quality and actually increase in value.

Though the program was sound practically, the Navajos did not agree with the reasoning behind it. To them, a large herd of sheep was a symbol of prestige and an integral part of the proper way to live. Administration of the stock reduction program did not take into account these important Navajo values. Government agents shot thousands of head of livestock and left their bodies to rot; the Navajo people were bewildered by the apparent waste. The program aroused resentment and hostility, and left them with a long-lasting distrust of the U.S. government and of Anglos in general.

Though herding is still widespread, importance of livestock in the economy has decreased in the past 40 years. The reduction of stock, along with experimental breeding programs, has helped improve the quality of Navajo sheep and increase the production of meat and wool. Other federal programs have aimed at improving Navajo agriculture by increasing the amount of land under cultivation through irrigation projects.

The size of the Navajo reservation has grown repeatedly; today it covers an area slightly larger than the states of Connecticut, Massachusetts, and New Hampshire combined. Still the barren land is not large enough to comfortably support the rapidly increasing population. Many people have left the reservation to live and work in border towns like Flagstaff, Arizona and Gallup, New Mexico. Others have moved even farther to big cities like Phoenix and Los Angeles, where they face such problems as poor housing and unemployment due to discrimination and lack of skills. Many find it difficult to adapt to the alien urban environment and culture.

Most Navajos continue to live on the reservation, where they feel at home with relatives and a familiar way of life. Because of the scarcity of jobs there, many must leave seasonally to earn money — working for the railroad, for example — to carry them through the rest of the year. The tribe and the federal government have in recent years tried to create more jobs on the reservation by attracting industries and encouraging establishment of tribally owned and operated businesses. But the average annual income for the Navajo is still far below the national average.

The federal government holds reservation lands "in trust" for the Navajo people. The Bureau of Indian Affairs is the branch of government responsible for administration of reservation matters. Navajos do not own land on the reservation in our sense of "private property," but a family usually occupies a traditional use area of grazing lands, water holes, farming plots, and camps that it has used for generations.

Though the federal government through the B.I.A. still has considerable financial and political authority over Navajo affairs, more and more the tribe is taking responsibility for its own people and their interests. The Navajo tribal government, headed by an elected chairman, has its headquarters in Window Rock, Arizona. Chapter houses represent tribal government on the local level. Each district elects a representative to the Tribal Council in Window Rock. Navajos also vote in state and federal elections.

Income from oil, coal, and uranium leases has in recent years added millions of dollars to the tribal treasury. The tribe maintains a police force, a system of tribal courts, a parks and recreation department, a fish and game department, and a tribal welfare department.

The visitor to the Navajo reservation today cannot help noticing the influence of modern American culture and technology on the traditional Navajo way of life. Though many still use horses for herding and for travel in remote areas, the pick-up truck has replaced the horse and the wagon as the primary means of transportation. People can now easily drive hundreds of miles to shop in Flagstaff and Gallup. Radio and even television form part of the daily consciousness of many Navajos. The People are no longer isolated from the white man's world.

Navajo men today wear cowboy garb. The older women still retain the colorful style of dress adopted at Fort Sumner, but most younger ones follow modern fashion trends. Though many Navajos still occupy traditional hogans, more and more live in square, Anglo-style frame houses or mobile homes—both on and off the reservation. Reservation towns like Tuba City, Kayenta, and Window Rock now offer Navajos such conveniences as supermarkets, banks, and hairdressers.

In spite of these modern intrusions, much of the traditional way of life still remains intact, especially in the more remote, western part of the reservation around Navajo National Monument. Many Navajos have no electricity or indoor plumbing; they haul water and firewood to their family camps, located miles from the pavement over rough dirt roads. They continue to hold the traditional ceremonies and dances; many

Today the pick-up truck enables Navajos to drive hundreds of miles to shop in towns like Flagstaff and Gallup. No longer are they isolated from the white man's world.

turn to the local medicine man to cure their illnesses. Traditional values still shape the world view of many Navajos, and they hold to many of the old customs. The slow and easy pace of reservation life contrasts with the bustle of the Anglo world around it.

"Traditional values still shape the world view of many Navajos, and they hold to many of the old customs."

Contacts between the two vastly different cultures — the Navajo and the Anglo — have not always been peaceful or easy. The two groups no longer clash in armed battle; today the battlefield is in the personal lives of the Navajo people. Some have been able to adapt successfully; but many others feel caught between the two cultures — unable to participate fully in either. Alcoholism, violence, and loss of identity are problems for many Navajos trapped in this limbo.

What will happen to the Navajo in the future? Certainly there are no simple answers to the problems facing the People today. Hopefully the resource fulness and adaptability demonstrated in the past will help them find a comfortable compromise between the lure of modern technological convenience and the special beauty of their traditional way of life. Disappearance of Navajo culture would be a tragic loss — for the People themselves, and also for the rest of us. The existence of their unique and different way of life and looking at the world adds richness to all of our lives.

SUGGESTED READINGS

Anderson, Susanne. 1973. *Song of the Earth Spirit.* Friends of the Earth, McGraw Hill Book Company, New York

Ambler, J. Richard. 1977. *The Anasazi.* Museum of Northern Arizona, Flagstaff, Arizona

Baars, Donald L. 1972. *Red Rock Country: The Geologic History of the Colorado Plateau.* Doubleday/Natural History Press, Garden City, New York

Dean, Jeffery S. 1969. *Chronological Analysis of Tsegi Phase Sites in Northeastern Arizona.* Papers of the Laboratory of Tree-Ring Research No. 3, University of Arizona Press, Tucson, Arizona

Downs, James F. 1972. *The Navajo.* Holt, Rinehart and Winston, Inc., New York

Dozier, Edward P. 1970. *The Pueblo Indians of North America.* Holt, Rinehart and Winston, Inc., New York

Hegmann, Elizabeth. 1963. *Navajo Trading Days.* University of New Mexico Press, Albuquerque

James, Harry C. 1974. *Pages from Hopi History.* University of Arizona Press, Tucson, Arizona

Hyde, Philip and Stephen C. Jett. 1969. *Navajo Wildlands: "as long as the rivers shall run."* Sierra Club, Ballantine Books, New York

Kluckhohn, Clyde and Dorothea Leighton. 1962. *The Navajo.* The Natural History Library, Anchor Books. Doubleday and Company, Inc., Garden City, New York

Longacre, William A., ed. 1970. *Reconstructing Prehistoric Pueblo Societies.* University of New Mexico Press, Albuquerque

Martin, Paul S. and Fred Plog. 1973. *The Archaeology of Arizona.* Doubleday/ Natural History Press, Garden City, New York

McNitt, Frank, 1957. *Richard Wetherill: Anasazi.* University of New Mexico Press, Albuquerque

Muench, David and Donald G. Pike. 1974. *Anasazi: Ancient People of the Rock.* American West Publishing Company, Palo Alto, California

Terrell, John U. 1970. *The Navajos: The Past and Present of a Great People.* Weybright and Talley, New York

Ward, Albert. 1975. *Inscription House.* Museum of Northern Arizona Technical Bulletin No. 16, The Northern Arizona Society of Science and Art, Flagstaff, Arizona

Waters, Frank. 1963. *Book of the Hopi.* Ballantine Books, New York

CREDITS

Artifacts p. 20, photo p. 57: Courtesy Arizona State Museum, Tucson.

Photos p. 51 bottom, p. 55 top and bottom: Courtesy Arizona State Historical Society, Tucson.

Artifact details cover and pp. 7, 15, 35, 63: by Tom Masamori, Courtesy Colorado Historical Society.

Photos pp. 5, 28, 65, 66, 67, 68: Courtesy Museum of Northern Arizona, Flagstaff.

Photo p. 31: Courtesy Milwaukee Public Museum, Milwaukee, Wisconsin.

Photo p. 56: Courtesy New Mexico State Records Center and Archives, Santa Fe—Frank McNitt Collection

Photo p. 51 top: Courtesy Laboratory of Tree-Ring Research, University of Arizona, Tucson.

Photo p. 23: by Keith Anderson, Courtesy of National Park Service.

Photos pp. 63, 73: by John Running.

Photo p. 61 right: by William Stoughton.

All other photos by author.

All maps by Pamela Lungé: prehistoric culture map on p. 17 after map in T. Weaver, *Indians of Arizona,* University of Arizona Press, 1974.